Advances in
High-Pressure Research

Advances in
High-Pressure Research

edited by

R. H. WENTORF, Jr

General Electric Company, Research and Development Center
Schenectady, New York, USA

Volume 4

1974

ACADEMIC PRESS
LONDON AND NEW YORK
A Subsidiary of Harcourt Brace Jovanovich, Publishers

ACADEMIC PRESS INC. (LONDON) LTD
24–28 Oval Road
London NW1

US edition published by
ACADEMIC PRESS INC.
111 Fifth Avenue,
New York, New York 10003

Library of Congress Catalog Card Number: 65-27317
ISBN: 0-12-021204-8

PRINTED IN GREAT BRITAIN BY
ADLARD AND SONS LTD
BARTHOLOMEW PRESS, DORKING

Contributors

W. A. Bassett, *Department of Geological Sciences and Space Science Center, University of Rochester, USA*

R. C. Crewdson, *Stanford Research Institute, California, USA*

D. R. Curran, *Stanford Research Institute, California, USA*

W. J. Murri, *Stanford Research Institute, California, USA*

C. F. Petersen, *Stanford Research Institute, California, USA*

T. Takahashi, *Department of Geological Sciences and Space Science Center, University of Rochester, USA*

R. H. Wentorf, Jr, *General Electric Company, Research and Development Center, Schenectady, New York, USA*

Contents

Contributors v

Response of Solids to Shock Waves
W. J. Murri, D. R. Curran, C. F. Petersen and R. C. Crewdson   1

X-ray Diffraction Studies up to 300 kbar
W. A. Bassett and T. Takahashi 165

Diamond Formation at High Pressures
R. H. Wentorf, Jr 249

Author Index 282

Subject Index 293

Response of Solids to Shock Waves

W. J. Murri, D. R. Curran, C. F. Petersen and
R. C. Crewdson[1]

Stanford Research Institute, California, USA

1	Introduction	2
2	Shock waves and their production in the laboratory	3
3	Constitutive relations	6
	3.1 Completely relaxed fluid	7
	3.2 Rate-independent elastic–plastic solid	8
	3.3 Materials with viscosity	10
	3.4 Stress-relaxing solid	12
	3.5 Dislocation models	14
	3.6 Porous solid	16
	3.7 Composites	22
	3.8 More general constitutive relations	24
4	Stress wave propagation	26
	4.1 Consequences of the conservation laws	26
	4.2 Consequences of the Hugoniot equations	29
	4.3 Unsteady wave solutions	36
	4.4 Current problems in wave propagation calculations	49
5	Instrumentation and measurement techniques	50
	5.1 Measurement of stress	51
	5.2 Particle velocity measurement	78
	5.3 Measurement of wave velocity	88
	5.4 Measurement of specific volume	89
	5.5 Measurement of shock temperature	90
	5.6 Experimental determination of constitutive relations	91
6	Effects of shock waves in solids	97
	6.1 Electrical effects	97
	6.2 Optical effects	112
	6.3 Phase transitions	115
	6.4 Magnetic properties	122

[1] Now at Industrial Noise Service Inc., Palo Alto, California.

6.5   Dynamic yielding   124

6.6   Residual effects   131

6.7   Shock-induced fracture   141

7   Comparison of static and dynamic results   143

References   146

1   Introduction

The study of shock waves in solids has changed its character significantly during the past several years. The early stage of development of the field began during World War II and was summarized in the review article by Rice *et al.* (1958). At that time the experimental measurements were mostly confined to measurements of shock velocities and the velocities of moving free surfaces, using electrical contact pins and high-speed cameras. These data were combined with fluid mechanical theory to obtain high-pressure thermodynamic equations of state in which the solids at extreme pressures were treated as fluids.

The next period of development was reviewed by Duvall and Fowles (1963) and consisted mainly of increased sophistication in the measurement of the same quantities: shock velocities and free surface velocities. However, several other techniques, such as the use of quartz gauges to measure stress at the sample–quartz interface, were developed during this period, and the region of lower stresses where material strength is important was explored. In addition, the development of computer codes for calculating wave propagation in elastic–plastic media was begun. At that time the equations of state that were used to describe the solids under shock loading were still reasonably simple, being mainly rate-independent, elastic–plastic relations.

Since then there have been several rather dramatic developments:

a. The extension of experimental time resolution into the nanosecond region.

b. The development of laser interferometry and in-material piezoresistive and electromagnetic gauges for measurement of stress and particle velocity histories in the interior of the material, and the development of analytical techniques to interpret such data.

c. The application of theoretical models of dislocation dynamics to relate macroscopic constitutive relations to microscopic material properties.

d. Precision measurements of a wide range of shock effects including shock-induced electrical polarization and anomalous thermoelectric effects.

e. The development of computer codes that can compute wave propagation in materials that are porous or composite, that may be undergoing phase changes, that have time-dependent elastic–plastic constitutive relations, and that may be shock loaded by processes ranging from simple impact to sudden in-depth radiation deposition.

The object of this paper is to describe these advances in some detail and to emphasize the change in viewpoint that has been brought about by the appearance of in-material gauges of high time resolution. Whereas previous reviews have emphasized the concept of the Hugoniot equation of state (which describes the thermodynamic states attainable through a *steady* shock wave), we shall emphasize the concept of *unsteady* waves and shall discuss the analytical procedures that have been developed to handle experimental data from unsteady waves in solid materials.

Several excellent review articles written during the last few years discuss many of the above subjects, and we have attempted in the present paper to avoid excessive overlap. Wherever practical, we have referred the reader to the earlier review and have concentrated on more recent developments.

2   Shock waves and their production in the laboratory

For the purpose of this paper we will use the term shock wave to mean a high-amplitude stress wave. For ease of measurement and theoretical interpretation of the measurements, it is generally desired to produce plane stress waves with one-dimensional strain. In stress regions for which compressibility decreases with increasing pressure, the compressive wavefront of such waves will tend to sharpen (shock-up) as the wave propagates until nonconservative forces counter this tendency and cause the formation of a sharp steady-state wavefront characterized by a shock velocity U dependent only on peak stress. (This is strictly true only if the stress is steady behind the wavefront. See Duvall and Fowles (1963).) The release or rarefaction front of such a pulse will continually spread out as propagation proceeds.

In the laboratory, one-dimensional shock waves are generally produced through the impact of a flat flyer plate against a stationary plate.

The specimen to be studied is usually the stationary plate itself or is attached to it. A schematic diagram illustrating the basic phenomena involved in such an impact is shown in Fig. 1. To simplify the discussion, the flyer and specimen are of the same material and are represented respectively by an infinite slab and a half space. Before impact the flyer was travelling at velocity ξ, and the specimen was at rest. Just after impact, as shown in Fig. 1, the material on either side of the impact interface has been compressed to a density ρ, raised to a stress σ, and

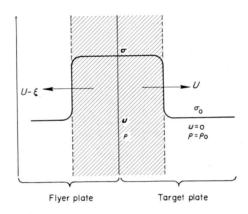

Fig. 1. Schematic of shock wave in target and flyer plates soon after impact.

accelerated to a velocity u, the "particle velocity". In Fig. 1 the stress as a function of distance is superimposed on the schematic and is shown propagating into flyer and specimen at a velocity U relative to the material. Upon reaching a free surface, in this case the back surface of the flyer, the shock is reflected as a rarefaction, which then propagates back into the material relieving the stress to zero as it goes. The end result of the process described in Fig. 1 is the production of a shock wave moving into the specimen with the idealized stress–distance profile shown in Fig. 2.

For most simple experiments we are interested in the magnitude of U, u, σ, and ρ as well as in the duration of the stress pulse itself. The first four are, as discussed in section 4, related by the conservation equations so that only two are independent and in need of measurement. In the simple example described above, the length of the stress pulse at the instant the shock reaches the rear surface of the flyer is simply twice the thickness of the flyer itself. Thus in this case the flyer thickness and the relative velocities of the shock and rarefaction are the controlling factors in the duration of the stress pulse at a point in the specimen. Since the flyer and specimen are not of infinite diameter, the upper limit of time duration of plane strain conditions at any point within the specimen is governed by the time it takes for rarefactions from the edge of the impact area to propagate into the point of interest. There is of course no sharp cut-off point, but for flyer plates less than about 2 in. in diameter, the time and space limitations on most experiments become severe. Stevens and Jones (1972) have discussed the design of experiments and the necessity of minimizing radial release effects.

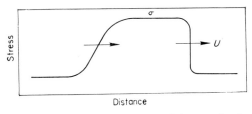

Fig. 2. Shock wave moving into target, followed by rarefaction unloading wave.

The methods commonly used to produce such plate impacts have been reviewed by Duvall and Fowles (1963) and Keeler and Royce (1971, pp. 51–80). We restrict ourselves here to summarizing the main impact techniques. The most widely used method for producing precisely controlled shock waves in the low to intermediate stress range is the light gas gun (Hughes et al., 1961; Thurnborg et al., 1964; Linde and Schmidt, 1966; Cable, 1970). In this method a flat-faced projectile, which is usually between 2 in. and 4 in. in diameter, but can be larger, is accelerated by compressed gas to velocities of up to 1.5 mm μs^{-1} and impacts a specimen attached to the gun muzzle. The plate on the front of the projectile can be any thickness down to about 0.01 in. This device is used principally in the 1 to 100 kbar range in most materials, and its main advantages are the availability of a continuous range of well-controlled velocities, precise control of angular misalignment between impacting surfaces (tilt), and convenience in use.

Another gun device of major interest is that reported by Jones et al. (1966). This is a modification of the light gas gun described above in that several stages of acceleration are used. This gun is capable of achieving projectile velocities of up to 8 mm μs^{-1}. Unfortunately, the small

(29-mm) projectile diameter of this gun places a restriction on the experiments that can be performed in uniaxial strain.

Another major technique suitable for producing shock pulses is the use of explosives to accelerate a flyer plate. In the simplest form of this method, an explosive plane wave generator (Duvall and Fowles, 1963) is used to detonate a pad of high explosive in contact with the flyer plate (often the pad is omitted). The method is most useful in the range of flyer plate velocities of from 1 to 5 mm μs^{-1}, although higher velocities have been reported (Al'tshuler *et al.*, 1958). Typically, flyer plates are several millimetres thick and are up to 12 in. in diameter. Other techniques for producing shocks include the use of explosives in contact with the specimen, high-intensity electron irradiation (Perry, 1970; Lutze, 1971; Bailey and Lutze, 1972), exploding foil methods (Cable, 1970), laser impulses (Jones, 1971), and magnetically driven flyer plates (Farber and Ciolkosz, 1969).

3   Constitutive relations

Up to now we have discussed stress waves in solids without discussing in mathematical detail how such stress waves arise. In principle the stress wave propagation can be predicted by solving the equations of conservation of mass, momentum, and energy subject to the appropriate initial and boundary conditions. However, the conservation equations are not sufficient in themselves; they must be supplemented by another equation that contains information about the specific material being loaded. This equation may specify a relationship between stress, strain, particle velocity, position, and time. The relation is sometimes simply called the specifying equation (von Mises, 1958), but it is also referred to as the stress–strain equation, the equation of state, or, particularly when the relation is time-dependent, the constitutive relation. Possibly the simplest example of such a relation is that for an incompressible material. In this case the relation becomes simply $\rho = $ constant. Another simple example is that of a linearly elastic solid for which the relation for the dependence of stress on strain is given by Hooke's law. Solids are often close to behaving in a linear-elastic manner at low stresses and for low rates of loading, but at the high stresses and rates attained in shock waves experiments most solids exhibit a complicated behaviour that is due to a more complicated constitutive relation. In this section we shall describe several constitutive relations of varying degrees of complexity and then discuss the connection between these macroscopic relations and microscopic processes such as dislocation behaviour. In a later section (4) we combine the constitutive relation with the conservation equations and discuss the problem of obtaining a solution for the resulting system of equations for the stress wave history.

3.1 COMPLETELY RELAXED FLUID

For stresses much higher than the yield stress, a solid behaves in the first approximation like a fluid, since the percentage deviations from an isotropic stress distribution are small. If, in addition, all time-dependent processes are completely relaxed, the constitutive relation will assume one of the forms of the equation of state for a non-reactive liquid or gas: $f(P, V, T) = 0$, $g(P, V, S) = 0$, $h(P, V, E) = 0$, $j(E, V, S)$ and so on. In short, one dependent thermodynamic variable can be written as a function of two independent thermodynamic variables. Of the above, only $(E-V-S)$ equations of state may be considered complete in the sense that they completely characterize a state (E, S, V, P, T). This topic as related to shock waves has been discussed by Cowperthwaite (1969).

A simple example of an equation of state is the ideal gas equation

$$P = (\gamma - 1) E/V, \tag{1}$$

where P is the pressure, E is the internal energy, V is the volume, and γ is the ratio of specific heats. Solids at high pressure, however, are too complicated to be well described by equation (1). Many high-pressure equations of state for relaxed solids have been proposed. Two that have seen much use are Murnaghan's equation (1937):

$$P = A[(\rho/\rho_0)^m - 1], \tag{2}$$

where ρ is the density and A and m are material parameters, and the Mie–Grüneisen equation (Grüneisen, 1926):

$$PV + G(V) = \Gamma(V) E, \tag{3}$$

where $\Gamma(V)$ is the Grüneisen coefficient for the material, and $G(V)$ is related to the lattice potential. If P_i, V_i, E_i and P_j, V_j, E_j are values of the parameters on the ith and jth thermodynamic paths, respectively, then subtraction of equation (3) for the two paths when $V_i = V_j = V$ yields

$$P_i - P_j = (\Gamma(V)/V) (E_i - E_j). \tag{4}$$

The ith and jth thermodynamic paths are arbitrary; for example, the ith path might be an isentrope and the jth path an isotherm.

The Murnaghan equation (2) has been extensively used in shock-wave physics but has no temperature dependence and is thus unsuitable for cases where shock heating is important. The Mie–Grüneisen equation (3) or (4) is based on lattice vibration theory and has become perhaps the most-used equation of state for high-pressure shock waves since applied by Walsh, Rice, McQueen, and Yarger in 1957 (Walsh *et al.*, 1957). Equation (4) is particularly convenient since the jth path may be taken to be a reference path, such as an experimentally deter-mined isotherm, and the ith path can be chosen as the path taken by the material in a shock wave. More traditionally, the reference path is taken to be the experimental locus of states attainable through a steady shock wave (the Hugoniot curve), and equation (4) is used to calculate isotherms and isentropes in the vicinity of the Hugoniot curve. The Grüneisen coefficient $\Gamma(V)$ has been the subject of much research and discussion (Walsh *et al.*, 1957; Rice *et al.*, 1958). Two expressions for $\Gamma(V)$ are often discussed: the Slater (1939) formula,

$$\Gamma = -\frac{V}{2}\left(\frac{\mathrm{d}^2 P/\mathrm{d}V^2}{\mathrm{d}P/\mathrm{d}V}\right) - \frac{2}{3},$$

and the Dugdale and MacDonald (1953) formula,

$$\Gamma = -\frac{V}{2}\left(\frac{\mathrm{d}^2 (PV^{2/3})/\mathrm{d}V^2}{\mathrm{d}(PV^{2/3})/\mathrm{d}V}\right) - \frac{1}{3}.$$

In many cases it can be simply assumed that either Γ or Γ/V is a constant (McQueen *et al.*, 1970), and in fact in practice Γ is often used as an adjustable parameter.

Comparison of equations (1) and (4) shows that the Mie–Grüneisen equation is similar in form to the equation of state of an ideal gas, with the role of the ideal gas specific heat ratio being played by $\Gamma + 1$ of the solid. This analogy has recently been discussed by Huang (1970).

3.2   RATE-INDEPENDENT ELASTIC–PLASTIC SOLID

Most solids exhibit elastic–plastic behaviour. That is, if the solid is compressed or distended slowly, the solid will obey Hooke's law of linear elasticity until a yield condition is reached. For higher loads the material will deform plastically. If this process does not depend upon the rate of loading, a fairly simple constitutive relation may be constructed. This problem applied to shock loading has been discussed by Morland (1959) and by Fowles (1961).

The approach is based on the assumption that an increment in the total strain ϵ_{ij}^{p} is the sum of the elastic strain increment and the plastic strain increment:

$$d\epsilon_{ij} = d\epsilon_{ij}^{e} + d\epsilon_{ij}^{p}. \tag{5}$$

The elastic strain increment is obtained from Hookes law:

$$d\epsilon_{ij}^{e} = (d\sigma_{ij}'/2\mu) + (1/3K)\,\delta_{ij}\,dP, \tag{6}$$

where μ is the shear modulus, K is the bulk modulus, $dP = (1/3)\,d\sigma_{ii}$, the deviator stress $\sigma_{ij}' \equiv \sigma_{ij} - P$, δ_{ij} is the Kronecker delta, and where a repeated index denotes summation. Thus, the stress is supposed to depend only upon the elastic part of the strain. Fowles (1961) included work hardening in the von Mises yield criterion,

$$\Upsilon = (3/2)^{1/2}(\sigma_{ij}'\sigma_{ij}')^{1/2}, \tag{7}$$

by letting $\Upsilon = \Upsilon(W_{p})$, where Υ is the yield stress in simple tension, and W_{p} is the plastic work defined by

$$dW_{p} = \sigma_{ij}\,d\epsilon_{ij}^{p}. \tag{8}$$

It is also assumed that the plastic deformation causes no volume changes, so that

$$d\epsilon_{ii}^{p} = 0. \tag{9}$$

Fowles combined equations (5) to (9) to derive the constitutive relations for the case of uniaxial strain:

Elastic region, $\sigma_x < [(1-v)/(1-2v)]\Upsilon^0$

$$\epsilon_y = \epsilon_z = 0$$
$$P \equiv \tfrac{1}{3}(\sigma_x + \sigma_y + \sigma_z) = K\,\epsilon_x$$
$$\sigma_x' \equiv \sigma_x - P = \tfrac{4}{3}\mu\epsilon_x \tag{10}$$
$$\sigma_y = \sigma_z = [v/(1-v)]\,\sigma_x.$$

Plastic region, $\sigma_x > [(1-v)/(1-2v)]\Upsilon^0$

$$\epsilon_y = \epsilon_z = 0$$
$$dP = K\,d\epsilon_x$$
$$d\sigma_x = dP + \tfrac{2}{3}\,d\Upsilon$$
$$d\epsilon_x^{p} = \tfrac{2}{3}[d\epsilon_x - (d\Upsilon/2\mu)] = -2d\epsilon_y^{p} \tag{11}$$
$$d\epsilon_x^{e} = \tfrac{1}{3}\,d\epsilon_x + (d\Upsilon/3\mu)$$
$$d\epsilon_y^{e} = \tfrac{1}{3}\,d\epsilon_x - (d\Upsilon/6\mu)$$
$$\sigma_x' \equiv \sigma_x - P = \tfrac{2}{3}\Upsilon$$

where ν is Poisson's ratio, $\varUpsilon^0$ is the initial value of the yield stress (taken positive in compression), and where ϵ^p and ϵ^e denote plastic and elastic components of strain, respectively. If there is no work hardening, $d\varUpsilon$ will be zero in the relations (11).

The above behaviour is shown schematically in Fig. 3. If the solid is compressed such that lateral movement at the boundaries is not allowed, the material will deform elastically until $\sigma_x = [(1-\nu)/(1-2\nu)]\varUpsilon^0$.

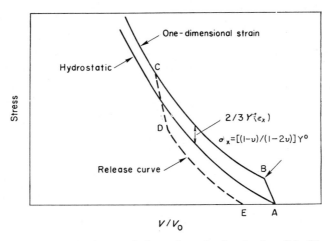

Fig. 3. Stress–strain paths in rate-independent elastic–plastic solid. (From Fowles, 1961.)

Thereafter, the deformation will follow a stress-volume path offset from the hydrostatic curve by an amount $\frac{2}{3}\varUpsilon(\epsilon_x)$. Upon release, the material will deform elastically until the stress is below the hydrostatic curve by $\frac{2}{3}\varUpsilon(\epsilon_x)$, after which it will deform plastically back to $\sigma_x = 0$. The material is left with a net strain due to the hysteresis, and since $|\sigma_x - P| = \frac{2}{3}\varUpsilon$, $P = \frac{2}{3}\varUpsilon$ at this point. Since $P = \frac{1}{3}(\sigma_x + \sigma_y + \sigma_z) = \frac{2}{3}\sigma_y$, an unrelaxed lateral stress $\sigma_y = \varUpsilon$ is still present because of the lateral constraints which prevent lateral expansion.

3.3 MATERIALS WITH VISCOSITY

All real solids that exhibit plasticity can also be expected to exhibit rate effects under dynamic loading. This is because the microscopic processes (such as dislocation motion) that cause plastic flow have characteristic relaxation times that govern the maximum rate at which the plastic flow can occur. Thus, if the loading time is of the same order as or smaller than the characteristic relaxation time for plastic flow, a rate effect should be present. In plate slap experiments the loading time at the projectile–target interface can be made almost vanishingly small; thus rate effects must be initially present. Whether such rate effects are observed experimentally depends on whether the experimental time resolution is smaller than the characteristic relaxation time of the material. In fact, such relaxation effects have been observed with plate slap experiments in many solids (Taylor, 1965; Gilman, 1968; Jones and Mote, 1969; J. N. Johnson et al., 1970).

It is therefore necessary to have constitutive relations that are time dependent. Probably the simplest such relation is that for a simple fluid that exhibits linear shear viscosity. In this case a viscous stress arises that is proportional to the shear strain rate:

$$\tau_{ij} = \eta \dot{\gamma}_{ij}, \tag{12}$$

where the shear stress $\tau_{ij} = \sigma_{ij}$, $i \neq j$, $\gamma_{ij} = 2\epsilon_{ij}$, and η is the shear viscosity coefficient.

Another simple example is that of a linear viscoelastic solid, where deviations of the stress from linear-elastic behaviour are proportional to strain rate:

$$\tau_{ij} = \mu \gamma_{ij} + \eta \dot{\gamma}_{ij}. \tag{13}$$

(More general viscoelastic relations will be discussed later.)

A constitutive relation more often used in shock wave work is the pair of elastic–viscoplastic relations:

$$\tau_{ij} = \mu \gamma_{ij}, |\tau| \leq Y/2, \text{ elastic} \tag{14}$$

$$\tau_{ij} = Y/2 + \eta \dot{\gamma}_{ij}, |\tau| > Y/2, \text{ viscoplastic.} \tag{15}$$

In the special case of uniaxial strain in a homogeneous, isotropic material where $\epsilon_{22} = \epsilon_{33} = 0$ and $\sigma_{22} = \sigma_{33}$, the maximum shear stress occurs at an angle of $45°$ from the σ_{11} direction, and

$$\tau_{\max} = (\sigma_{11} - \sigma_{22})/2.$$

Also,

$$\gamma_{\max} = \epsilon_{11} - \epsilon_{22}.$$

Define

$$\theta \equiv \epsilon_{11} + \epsilon_{22} + \epsilon_{33} = \epsilon_{11} + 2\epsilon_{22}$$

and recall that

$$P \equiv \tfrac{1}{3}(\sigma_{11} + \sigma_{22} + \sigma_{33}) = \tfrac{1}{3}(\sigma_{11} + 2\sigma_{22}).$$

The combination of these equations with equation (15) yields

$$\sigma_{11} - P = \tfrac{2}{3}\varUpsilon + 2\eta(\dot{\epsilon}_{11} - \dot{\theta}/3).$$ (16)

Since $\epsilon_{22} = 0$, $\dot{\theta} = \dot{\epsilon}_{11}$, and equation (16) becomes

$$\sigma' = \sigma_{11} - P = \tfrac{2}{3}\varUpsilon - \tfrac{4}{3}\eta\dot{\epsilon}_{11}.$$ (17)

In the above derivation we have assumed that γ is the total strain and is equal to the sum of the elastic and plastic strains. It could be argued that a more realistic physical model would allow the viscous shear stress to depend only on the plastic component of strain rate. In this case, equation (15) would be replaced by

$$\tau_{ij} = \varUpsilon/2 + \eta\dot{\gamma}_{ij}^{\mathrm{p}}.$$ (18)

The argument would carry through as before, with equation (16) replaced by

$$\sigma_{11} - P = \tfrac{2}{3}\varUpsilon + 2\eta(\dot{\epsilon}_{11}^{\mathrm{p}} - \dot{\theta}^{\mathrm{p}}/3).$$ (19)

Assume as before that plastic deformation causes no volume change. Thus, $\dot{\theta}^{\mathrm{p}} = 0$, and we have

$$\sigma' = \sigma_{11} - P = \tfrac{2}{3}\varUpsilon + 2\eta\dot{\epsilon}_{11}^{\mathrm{p}}.$$ (20)

We see that the result of either equation (17) or (20) is that the material will deform much like the elastic–plastic solid shown in Fig. 3 with a $\tfrac{2}{3}\varUpsilon$ offset from the hydrostatic curve, but with an additional offset proportional to strain rate.

3.4   STRESS-RELAXING SOLID

As will be discussed later in the section on stress wave propagation, a limitation of the elastic–viscoplastic solid discussed above is that equation (17) or (20) has a term in $\dot{\epsilon}$ but none in $\dot{\sigma}$. This lack of symmetry is related to the fact that when equation (17) is added to the equations of conservation of mass, momentum, and energy, the resulting system of equations is parabolic instead of hyperbolic. A property of a parabolic system is that arbitrary initial values cannot be specified, since the $t = 0$ axis coincides with characteristic directions. In short, equation (17) is not suited for description of transient behaviour. An indication of this can be seen from equation (17) if we assume that the material is held at $\sigma_{11} = P + \tfrac{2}{3}\varUpsilon$, and then a constant strain rate $\dot{\epsilon}_{11}$ is suddenly applied. According to equation (17) the stress would then discontinuously jump by $\frac{4}{3}\eta\dot{\epsilon}_{11}$. This step-function increase is physically unrealistic. On the other hand, if equation (17) had included a term proportional to $\dot{\sigma}_{11}$, the stress would have responded continuously according to the solution of the differential equation for σ_{11}.

For the above reasons, constitutive relations that contain both $\dot{\sigma}$ and $\dot{\epsilon}$ terms are usually used in shock wave problems where transient effects are important. An easy way to arrive at such a relation is to again postulate a physical model in which the total strain in the direction of wave propagation is the sum of elastic and plastic components:

$$\epsilon_{11} = \overset{e}{\epsilon}_{11} + \overset{p}{\epsilon}_{11},$$

and

$$\dot{\epsilon}_{11} = \overset{e}{\dot{\epsilon}}_{11} + \overset{p}{\dot{\epsilon}}_{11}.$$

Now assume that $\epsilon_{22} = \epsilon_{33} = 0$ (uniaxial strain) and that only elastic strain gives rise to stress. Thus, from Hooke's law for $\overset{e}{\dot{\epsilon}}_{11}$,

$$\dot{\epsilon}_{11} = \dot{\sigma}_{11}/(K + \tfrac{4}{3}\mu) + \overset{p}{\dot{\epsilon}}_{11}. \tag{21}$$

Next assume that the plastic strain rate is a function of σ_{11}, ϵ_{11}, and t:

$$\overset{p}{\dot{\epsilon}}_{11} = G(\sigma_{11},\ \epsilon_{11},\ t). \tag{22}$$

Thus,

$$\dot{\epsilon}_{11} = \dot{\sigma}_{11}/(K + \tfrac{4}{3}\mu) + G(\sigma_{11},\ \epsilon_{11},\ t). \tag{23}$$

This relation for a stress-relaxing solid was discussed in detail for elastic–plastic wave propagation problems by Malvern (1951).

A simple form of equation (23) is obtained by setting

$$G(\sigma_{11},\ \epsilon_{11},\ t) = \frac{\sigma_{11} - (P + \tfrac{2}{3}Y)}{(K + \tfrac{4}{3}\mu)\ T}, \qquad \sigma_{11}' > \tfrac{2}{3}Y, \tag{24}$$

where T is a relaxation time. Combining equations (23) and (24) yields

$$\sigma_{11} - P = \tfrac{2}{3}Y + [(K + \tfrac{4}{3}\mu)\ T]\dot{\epsilon}_{11} - T\dot{\sigma}_{11}, \tag{25}$$

the equation for a Maxwellian solid. Comparison of this relation with equation (17) shows that if the deviator stress σ_{11}' is held constant, equations (25) and (17) become identical if μT is set equal to the shear viscosity η. Thus the constitutive relation for this stress-relaxing solid approaches that of an elastic–viscoplastic solid as the rate of deviator stress approaches zero.

Other similar models have been reported in which the yield stress Y

is also allowed to be rate dependent, thus providing two simultaneous relaxation processes (Seaman *et al.*, 1969).

3.5   DISLOCATION MODELS

Plastic deformation in metals is caused by the movement of dislocations. In a typical metal there are 10^6 to 10^9 centimetres of dislocation lines in a cubic centimetre of material; thus the large number of dislocations present makes a statistical averaging of individual dislocation motions an attractive method for relating macroscopic plastic deformation to the microscopic material properties. This approach was pioneered in papers by Gilman (1960, 1965), Johnston (1962), Gillis and Gilman (1965), Gillis *et al.* (1969), and was first applied to explain data from shock wave plate-slap experiments by Taylor (1965). During the last few years significant advances have been made in understanding dynamic crystalline material behaviour in the light of dislocation processes. Fairly recent reviews have been given by Gilman (1968), and by McQueen *et al.* (1970).

The basis of all dislocation models of plasticity is the Orowan equation for the plastic shear strain rate

$$\dot{\gamma}^{\mathrm{p}} = Nbv, \tag{26}$$

where N is the mobile dislocation density, b is the Burger's vector, and v is the *average* dislocation velocity. Both N and v may be functions of macroscopic stress and plastic strain. Thus an appropriate constitutive relation is obtained by setting $\dot{\gamma}^{\mathrm{p}}$ from equation (26) into the stress relaxation model equation (23) for $G(\sigma_{11}, \epsilon_{11}, t)$. For example, for the case of uniaxial strain, $\epsilon_{22} = \epsilon_{33} = 0$, and if the material can be considered homogeneous and isotropic, the maximum shear stress will occur at $45°$ to σ_{11}. Thus, $\tau_{\max}$ and $\gamma_{\max}$ are given by

$$\tau_{\max} = (\sigma_{11} - \sigma_{22})/2 = \tfrac{3}{4}\sigma'_{11}.$$

and

$$\gamma^{\mathrm{p}}_{\max} = \epsilon^{\mathrm{p}}_{11}.$$

Therefore,

$$G(\sigma_{11}, \epsilon_{11}, t) = \dot{\epsilon}^{\mathrm{p}}_{11} = \dot{\gamma}^{\mathrm{p}}_{\max} = N(\sigma'_{11}, \epsilon^{\mathrm{p}}_{11}) \, bv(\sigma'_{11}, \epsilon^{\mathrm{p}}_{11}),$$

and equation (23) becomes

$$\dot{\epsilon}_{11} = \dot{\sigma}_{11}/(K + \tfrac{4}{3}\mu) + N(\sigma'_{11}, \epsilon^{\mathrm{p}}_{11}) \, bv(\sigma'_{11}, \epsilon^{\mathrm{p}}_{11}). \tag{27}$$

Equation (27) is thus a possible constitutive equation for a crystalline solid loaded in uniaxial strain. The dislocation density function $\mathcal{N}(\sigma'_{11}, \epsilon^{p}_{11})$ and the average dislocation velocity function $v(\sigma'_{11}, \epsilon^{p}_{11})$ must be determined experimentally for each material.

An approximation made in the derivation of equation (27) is that the material can be considered homogeneous and isotropic so that the maximum shear strain occurs at 45° to the σ_{11} direction. This assumption has usually been justified for polycrystalline solids by the plausibility argument that enough crystals will have their slip planes oriented at 45° to σ_{11} so that most of the plastic shear deformation will in fact occur at this angle. A detailed examination of this problem by Johnson (1968) indicated that this assumption may lead to appreciable errors. However, in the case of single crystals loaded in uniaxial strain, the true orientations of the slip planes are known, and a more exact determination of $G(\sigma_{11}, \epsilon_{11}, t)$ can then be made (Jones and Mote, 1969; J. N. Johnson et al., 1970).

The dislocation model of the constitutive equation thus rests upon determination of the function $\mathcal{N}$ and v. As discussed by Kelly and Gillis (1967), and by Gillis et al. (1969), various differing forms for the function $v(\gamma^{p}, \tau)$ have been used. A functional form commonly used is the stress-activated relation of Gilman:

$$v = v^* \exp(-D/\tau), \tag{28}$$

where v^* is a limiting velocity and D is a "drag stress". Another possible form is the linear viscosity or dislocation damping model:

$$v = b\tau/B \tag{29}$$

where b is the Burger's vector and B is the viscous damping coefficient. A combination of equations (28) and (29) was suggested by Gillis et al. (1969):

$$v^*/v = \exp[1/a] + ma(v/c)/[(1 + m^2 a^2)^{1/2} - 1], \tag{30}$$

where $m = 2bD/cB$, $a = \tau/D$, and $c =$ shear wave velocity. Still another form is the thermal activation model (Kelly and Gillis, 1967), which replaces equation (26) by

$$\dot{\gamma}^{p} = \dot{\gamma}_0 \exp(-H/kT)\{\exp[V^*(\tau - \tau_g)/kT] - 1\}, \tag{31}$$

where $\dot{\gamma}_0$ is a frequency factor, H is the activation energy, V^* is the activation volume, and τ_g is a back stress.

At low stresses the dislocations cannot move past the short-range obstacles to their motion without the assistance of thermal agitation. In this region equations such as equation (31) are applicable. At higher stresses the dislocations can move without the aid of thermal agitation and it is often found that the viscous drag relation (29) describes the average dislocation velocity. At still higher stresses it appears that the dislocation velocity approaches a limiting velocity of the order of the shear wave velocity, and relations like equation (28) or equation (30) apply.

The dependence of the mobile dislocation density N on stress and plastic strain has been discussed in the review by Gilman (1968). In some cases N can be taken simply as a linear function of plastic strain. A more general relation, which takes strain hardening into account, is

$$N = (N_0 + M\gamma^{\mathrm{p}}) \exp \left[-\frac{H\gamma^{\mathrm{P}}}{\tau} \right], \qquad (32)$$

where M is a "multiplication coefficient" of the material, and H is a "hardening coefficient".

The varying degrees of success that have been attained in applying these dislocation models to the prediction of stress-wave histories in shock-loaded solids will be discussed in sections 4.3 and 6.5. In general, there is still a large amount of uncertainty in both the N and v functions, and current attempts to predict wave propagation phenomena from dislocation properties are to a great extent educated curve fitting. Nevertheless, the above approach is undoubtedly correct; the main problem is the difficulty of microscopically measuring independently the mobile dislocation densities and mobilities. Once these functions are available, it will be possible to derive macroscopic constitutive relations for crystalline solids from microscopic material properties, and this will form a powerful tool for predicting dynamic material behaviour.

3.6 POROUS SOLIDS

The use of porous solids (also called distended solids, or foams) has been found to be an effective means of protecting structures from shock damage caused by impulsive loads. This is because the peak stress of a propagating wave may be caused to attentuate very rapidly when it propagates through a porous solid. An unusual example is the recently reported effect of porosity on seismic data recorded on the surface of the moon (Latham *et al.*, 1970). To understand and to predict this behaviour, constitutive relations for porous solids must be known and much effort has been spent in recent years to develop such relations. The main distinguishing feature of a porous solid is that it exhibits "crushing", i.e. an irreversible collapse of the voids during compression. The qualitative quasi-static loading behaviour of a porous solid is

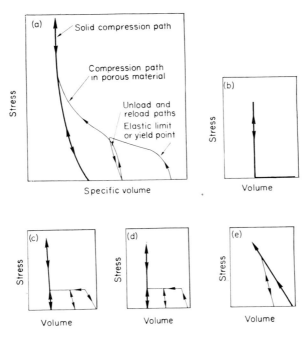

Fig. 4. Compression path of a porous material and four simple models for porous materials. (a) Compression path of porous material. (b) Simple locking or snowplow model. (c) Plastic–perfectly plastic rigid locking model. (d) Elastic–perfectly plastic locking model. (e) Elastic locking model. (From Seaman and Linde, 1969.)

illustrated in Fig. 4(a). In Fig. 4(b–e) several simple models that approximate this behaviour are shown. In general, a porous solid will exhibit elastic behaviour until the material begins to crush and to fill the pores. The static stress–strain curve will be similar to that of an elastic–plastic solid in this region. As the pores are filled in, the material approaches full compaction, and the compression path approaches that of the solid material. Various model equations of state that exhibit this behaviour have been developed by Brazhnik *et al.* (1962), Thouvenin (1965), Heyda (1969), Prindle (1968), Hofmann *et al.* (1968), Herrmann (1969a), Seaman and Linde (1969), Pastine *et al.* (1970), and Seaman (1970a).

A wealth of shock-wave experiments have been performed on porous solids. Various porous soils and rocks have been investigated by Ahrens and Gregson (1964), Anderson *et al.* (1966), Petersen (1969), and Lysne (1970a). Kormer *et al.* (1962) studied porous aluminium, copper, lead, and nickel, and Krupnikov *et al.* (1962) investigated the shock properties of porous tungsten. Porous graphite and aluminium were studied by Linde and Schmidt (1966a); porous iron by Lysne and Halpin (1968) and by Butcher and Karnes (1968); porous graphite, copper and tungsten by Boade (1968, 1969, 1970); porous carbon by Butcher (1966) and by Lysne (1970b); and porous magnesium and ammonium sulphate by Johnson and Wackerle (1967). Other porous materials studied include polyurethane, beryllium, and silica (Rempel *et al.*, 1966); celotex, styrofoam, redwood, sugar pine, foam glass, and balsa (Liebermann, 1967); boron nitride (Lee, 1967), and titanium dioxide (Linde and De Carli, 1969). Porous beryllium was studied by Eden and Smith (1970), porous copper, iron, tungsten, and polyurethane by Linde *et al.* (1972), and porous aluminium and polyurethane by Butcher *et al.* (1972). In addition, heated porous aluminium has been investigated by Ng and Butcher (1972), and distended carbon felt was studied by Lee (1972).

A model equation of state or constitutive relation that has found extensive use in describing experimental data is the $P-\alpha$ model developed by Herrmann (1969a). In the original version of this model, shear strength effects and rate effects were neglected so that an equation of state relating pressure P, specific volume V and specific internal energy E is sought:

$$P=f(V, E). \tag{33}$$

The surface energy of the pores is neglected so that the specific internal energy is the same for the porous material as for the solid material for given values of P and T. The degree of compaction α is defined as the ratio of the specific volume of the porous solid to the specific volume of the corresponding solid material at the same pressure and temperature:

$$\alpha = V/V_{s}. \tag{34}$$

If the rate-independent equation of state of the solid material, neglecting shear strength, is

$$P = f(V_s, E) \tag{35}$$

then the above assumptions allow the equation of state of the porous material to be written as

$$P = f(V/\alpha, E), \tag{36}$$

where f is the same function in the above two equations, i.e. the equation of state of the solid component of the material is independent of whether pores are present or not, and the equation of state of the porous material thus results from simply replacing V_s by V/α in the solid equation of state. If the degree of compaction α is a known function of the thermodynamic state,

$$\alpha = g(P, E), \tag{37}$$

then equations (36) and (37) constitute a complete equation of state for the porous material. Herrmann (1969a) points out that $g(P, E)$ could in principle be determined for a given porous material by performing a sufficient number of experiments. However, such data are at present not available, and the usual procedure is to approximate $\alpha = g(P, E)$ by $\alpha = g(P)$, where $g(P)$ is determined from shock compaction experiments on a given porous material initially at ambient pressure and temperature. As will be shown in section 4, in such experiments the pressure and internal energy are related by the nature of the loading process, so the dependence of α on E is implicit, and E is no longer an independent thermodynamic variable. Nevertheless, for experiments in which the internal energies are expected to be close to those required by the above restriction, the model has been successful in describing and predicting experimental results.

The $P - \alpha$ model has been extended to include rate effects by Johnson (1968) and by Butcher (1970). Butcher takes the time derivative of equation (34) to obtain

$$\dot{\alpha}/\alpha = \dot{V}/V - \dot{V}_s/V_s. \tag{38}$$

The next step is to decompose $\dot{\alpha}$ into an elastic part $\dot{\alpha}_e$ plus an irreversible (plastic) part $\dot{\alpha}_p$,

$$\dot{\alpha} = \dot{\alpha}_e + \dot{\alpha}_p. \tag{39}$$

By analogy to the stress-relaxing model of plasticity discussed earlier, $\dot{\alpha}_p$ is assumed to be a function of the thermodynamic variables:

$$\dot{\alpha}_p = G(P, E, \ldots). \tag{40}$$

The constitutive relation is obtained by combining equations (38) to (40):

$$\dot{V}/V = \left[\frac{\dot{\alpha}_e + G(P, E, \ldots)}{\alpha}\right] + \dot{V}_s/V_s. \qquad (41)$$

It is convenient to express the time-dependent terms on the right-hand side of equation (41) as functions of $\dot{P}$. The term $\dot{V}_s/V_s$ can be converted to terms in $\dot{P}$ by taking the time derivative of equation (35):

$$\dot{P} = (\partial f/\partial V_s)_E \, \dot{V}_s + (\partial f/\partial E)_{V_s} \, \dot{E}.$$

Since pore compaction is assumed to be adiabatic, the first law of thermodynamics yields

$$\dot{E} = -P\dot{V}.$$

Combining the above two relations yields

$$\dot{V}_s/V_s = [\dot{P} - PV(\partial f/\partial E)_{V_s} (\dot{V}/V)]/V_s(\partial f/\partial V_s)_E. \qquad (42)$$

In addition, Butcher assumes α_e to be a function of the thermodynamic variables P and E so that $\dot{\alpha}_e$ may similarly be expressed as a function of $\dot{P}$ and $\dot{V}$. Thus, for a given value of $\dot{V}/V$, the constitutive relation (41) can be written as a differential equation describing the time variation of P. Butcher assumed a function for $\dot{\alpha}_p = G(P, E)$ of the Maxwellian type (the same form as that in equation (24) for a stress-relaxing solid) and applied the resulting constitutive relation to describe the results of shock compaction experiments on polyurethane foam.

Although the P-α model and the time-dependent extensions of that model are in principle capable of handling the specific internal energy E as an independent thermodynamic variable, in practice they have been restricted to cases where E is determined by the nature of idealized shock loading processes, i.e. to values of E attainable through steady idealized shock waves. A recent attempt to construct rate-independent equations of state for porous material free of this restriction was recently reported by Seaman and Linde (1969) and by Seaman (1970a). The equation of state is represented by surfaces in E–V–P space as shown in Fig. 5. An interesting feature, which can be seen in Fig. 5, is that heating the partially compacted material at constant volume may cause the pressure to decrease to zero as the solid material melts and flows into the pores. Further heating at constant volume will cause the solid material to expand to full compaction, after which the pressure will increase with further heating. Such thermodynamic states may be attained by in-depth radiation deposition, but are far from those attainable through a steady shock wave originating at ambient conditions and thus would not be attainable with present forms of the P-α model (although the surfaces shown in Fig. 5 could in principle be generated by proper specification of $\alpha = g(P, E)$). Seaman and Linde (1969) used experimental results from impact experiments, material

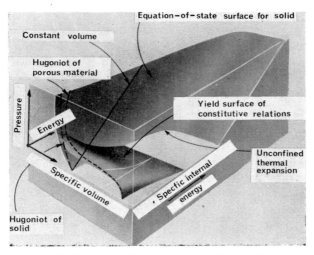

Fig. 5. Depiction in E–V–P space of the constitutive relations for a porous material. (From Seaman and Linde, 1969.)

parameters such as yield strength, spall strength, bulk moduli, and thermal expansion (plus a certain amount of intuition) to construct the surfaces shown in Fig. 5 for several porous metals. The behaviour of these materials when exposed to impact or nuclear radiation was then calculated, based on the constructed equations of state.

The brief discussion above indicates that the construction of constitutive relations for porous solids is still in the development stage and that much work remains to be done. Of particular interest is the need to develop a microscopic theory that would predict the macroscopic behaviour of a porous solid much as dislocation theory is currently used to derive constitutive relations for elastic–plastic solids. Efforts have recently been made in this direction by Warren (1973), Carroll and Holt (1972), Cowin (1972), and by Seaman et al. (1971).

3.7 COMPOSITES

The constitutive relations that we have discussed thus far have all been for one-component, homogeneous solids. However, many solid materials now used in engineering practice are composites made by bonding metal or quartz fibres in a plastic matrix or by bonding metal to metal, metal to ceramic, or plastic to ceramic. The advantage to be gained from such composites is that it is possible to attain otherwise hard-to-get combinations of properties. Examples include metal–ceramic composite armours which combine high hardness with ductility; re-entry vehicle heat shields, which combine high strength with low density and low thermal conductivity; and modern alpine skis, which are metal, wood, and polymer composites combining light weight with complex stiffness requirements. In addition, many naturally occurring materials such as rocks and wood are composites.

The rapid growth of the use of composites in engineering structures, as well as geophysical interest in wave propagation in rocks and soils, have made it important to characterize the response of composites to dynamic loads, and this requires the construction of appropriate constitutive relations. If the constitutive relations of the individual components are known, it is often possible to estimate the "average" constitutive relation of the composite, by forming weighted averages of the thermodynamic variables of the components and then placing these average variables into a constitutive relation such as the Mie–Grüneisen equation (4). This procedure has been discussed in the review article by McQueen et al. (1970) (see also Duvall and Taylor, 1971).

Similar attempts to construct average constitutive relations for composites have been reported by Tsou and Chou (1969, 1970), Chou and Wang (1970), Torvik (1970), Hopkins (1970), Lysne (1970a), Lundergan and Drumheller (1970, 1971, 1972), Barker (1971a), Munson and Schuler (1971), and Davis and Wu (1972). Since the approach used in the works cited above is to develop an "average" equation of state, the composite is treated as though it were a homogeneous material. This means that attenuation mechanisms such as geometric dispersion due to the heterogeneity of the composite are either ignored or possibly lumped together as effects of a material "viscosity". However, viscosity involves energy dissipation, which, in the purely elastic case, is inconsistent, and the question of the relative importance of these dispersive effects is still unresolved. Davis and Cottrell (1971) have extended the control-volume approach of Tsou and Chou to include chemical reactions among the constituents and phase transformations of a constituent.

An alternate approach is to attempt to analyse the interaction of the individual composite components in detail and to base wave propagation predictions upon the constitutive relations of the individual components. Such an approach is practical for composites of simple geometry when each component is treated as an elastic solid. Even with these simplifications, however, the problem is formidable, and a large body of literature exists concerning this approach. Recent review articles by Peck (1970, 1972) summarize the progress that has been made in this direction (see also Garg and Kirsch, 1971, 1973). One of the general results from this elastic approach is that the geometric dispersion is large whenever the characteristic wavelength of the pulse is of the same order or smaller than the scale of the composite heterogeneity. Such behaviour is intuitively reasonable. However, the theoretical and experimental work cited above by Lundergan and Drumheller, as well as recent work by Barbee *et al.* (1970) and by Seaman (1970a), indicate that the geometric dispersion is less than expected from elastic theory, that it all takes place within the distance defined by the first heterogeneity interface, and that the stress wave maintains a constant shape for greater distances into the composite material. These effects are probably caused by the importance of material porosity in these experiments (Seaman, 1970a) and by stresses being high enough to force at least one of the components into plastic flow, i.e. the constitutive relation would be nonlinear. In the work by Lundergan and Drumheller (1970), the composite consisted of alternate layers of 304 stainless steel and polymethyl methacrylate (PMMA), and the shape of a transmitted pulse was measured experimentally. The experimental results were then compared with calculations based on purely elastic constitutive relations and with calculations based on an elastic relation for the steel but with a hydrodynamic, nonlinear equation of state for the PMMA. The purely elastic calculations showed more geometric dispersion than was experimentally observed, whereas the nonlinear calculations showed good agreement with experiment.

In summary, the basic question is whether an "average" constitutive relation for a composite has meaning, i.e. whether it is necessary to calculate the local stress variation from component to component in a composite or whether it is possible or useful to define an "average" stress derived from an "average" constitutive relation. The answers to such questions will depend on the specific application and on the ratio of pulse length to composite scale. It is clear that much work remains to be done to develop useful constitutive relations that apply to the variety of problems concerned with dynamic loads in composite materials.

3.8 MORE GENERAL CONSTITUTIVE RELATIONS

The constitutive relations that we have discussed thus far have all been relatively simple. There is, however, a large body of literature in continuum mechanics dealing with more general forms of constitutive relations and their application to wave propagation. A detailed discussion of this subject is beyond the scope of this review, but we shall briefly mention two topics that are currently receiving much attention in connection with large amplitude wave propagation: materials with memory and finite-strain elastic–plastic theory.

In rate-independent constitutive relations the stress is given as a function of the instantaneous values of the strains. Thus, in these cases, the stress does not depend on the past history of the material. However, in the rate-dependent constitutive relations such as equation (23) for a Maxwell solid, the stress depends on material history, since integration of the relation will give stress as a function of the integral of a function of stress and strain over time. In general, a material for which the stress at a given time is determined by the previous history of the strain is called a "simple material". If the stress is much more dependent upon recent strain history than upon long-past events, the material may be classified as having "fading memory". It can be shown that linearly viscous fluids and viscoelastic materials are special cases of such materials. These concepts are discussed in detail in the treatise by Truesdell and Noll (1965). The thermodynamics of such materials and the propagation of disturbances in them have been handled in a series of papers by Coleman (1964), Coleman and Gurtin (1965a, 1965b, 1966), Coleman et al. (1965, 1966), and by Greenberg (1967), Gurtin (1968), and Chen and Gurtin (1970). An excellent recent review has been written by Herrmann and Nunziato (1973). These papers have established, for example, the general conditions under which steady shock waves can form in materials of this class. An interesting result in the paper by Chen and Gurtin (1970) is that a critical strain gradient behind a shock discontinuity exists for compressive shock waves in material with memory whose constitutive relation reduces to a linear viscoelastic equation at small strains. For strain gradients greater than the critical value, the shock wave amplitude will grow in time, whereas for values less than the critical value, the shock will decay. Thus, a steady shock will have a strain gradient equal to the critical value. The value of this critical strain gradient has been measured in polymethyl methacrylate by Schuler and Walsh (1970), and the result is in good agreement with the theoretical value.

In our discussion of constitutive relations for materials that exhibit plasticity we have assumed that the total strain is the sum of the elastic and plastic components of the strain. This is strictly true only for infinitesimal strains, whereas under shock conditions the elastic and plastic strains may both be finite. The problem of finite plastic strain has previously been handled by treating the plastic deformation as though it occurred in a fluid under hydrostatic stress, and the methods of fluid mechanics were then applied. The problem of treating the finite elastic strain remains, however, and until recently this has been routinely treated in shock wave calculations by infinitesimal strain theory. More recently, finite-strain elastic–plastic continuum models that can be applied to wave propagation problems have been developed by Backman (1964), Green and Naghdi (1965), Lee (1966), Lee and Liu (1967), Lee and Wierzbicki (1967), Herrmann (1969a), and Foltz and Grace (1969). These methods are generally based on expanding the specific internal energy function to third-order terms or higher in elastic strains and entropy and on finding a meaningful way to define elastic and plastic strains. The approach by Lee and collaborators has also recently been applied by Kelly (1970) to calculate steady shock structure. In addition, Clifton (1970a) used an approach similar to that of Lee to calculate the complete stress histories in one-dimensional strain, plate-impact experiments. Clifton found that the finite strain model yielded significantly different results than the infinitesimal strain model.

We may summarize as follows. Everything of interest relating to wave propagation that we know about the specific material is put into the constitutive relation for that material. As our measurements become more precise and we obtain better time and space resolution in our measurements, we need more detailed constitutive relations to describe

the material response. It is therefore to be expected that generaliza-
tions similar to those discussed above will become more important in
the future.

4　Stress wave propagation

4.1　CONSEQUENCES OF THE CONSERVATION LAWS

4.1.1　*One-dimensional strain waves*

As instrumentation has gradually improved in recent years, it has
become increasingly apparent that the simple assumptions (such as
steady-state propagation of a shock front and isentropic stress release in
the rarefaction) formerly used to analyse shock experiments are often if
not usually inaccurate. In this section, following Fowles and Williams
(1970), relationships are derived between the stress wave parameters
U, u, ρ, and σ general enough to apply to the most complex stress
waves actually observed in practice in uniaxial strain experiments. The
results are presented in the form used in practice to evaluate measurements made with stress and particle velocity gauges.

The conservation equations applied to wave propagation in a material
in which radiation,[1] heat conduction,[2] and other energy sources and
sinks are ignored are:

$$\left(\frac{\partial \sigma}{\partial x}\right)_t = -\rho \left.\frac{\mathrm{d}u}{\mathrm{d}t}\right|_p \tag{43}$$

$$\left.\frac{\mathrm{d}\rho}{\mathrm{d}t}\right|_p = -\rho \left(\frac{\partial u}{\partial x}\right)_t \tag{44}$$

$$\left.\frac{\mathrm{d}E}{\mathrm{d}t}\right|_p = -\frac{\sigma}{\rho} \left(\frac{\partial u}{\partial x}\right)_t \tag{45}$$

where

$$\left.\frac{\mathrm{d}}{\mathrm{d}t}\right|_p$$

[1] This must be taken into account if radiation deposition in depth is causing the shock.
Radiation heating or cooling can certainly be ignored in the centre (plain strain portion) of
an opaque specimen.
[2] Neglecting heat conduction has been shown to be valid except for shock wave "widths"
significantly narrower than those normally observed in practice (Cowan, 1965). In this
article, shock fronts and rarefractions are assumed to be adiabatic unless otherwise stated.

is here defined as the directional derivative along a particle path, i.e. a line of velocity u. Equations (43), (44), and (45) are, respectively, the equations of motion, continuity, and energy conservation written in laboratory (Eulerian) coordinates where x is the position in the direction of propagation and E is specific internal energy.

For an observer travelling at velocity ξ, the change of stress with time will be given by the directional derivative along a line of velocity ξ:

$$\frac{d\sigma}{dt}\bigg|_{\xi} = \left(\frac{\partial\sigma}{\partial t}\right)_{x} + \xi\left(\frac{\partial\sigma}{\partial x}\right)_{t}. \tag{46}$$

If the "observer" is a "Lagrangian" transducer imbedded in the material and travelling at particle velocity u, then we have the change of stress with time on a directional derivative along a "particle path":

$$\frac{d\sigma}{dt}\bigg|_{p} = \left(\frac{\partial\sigma}{\partial t}\right)_{x} + u\left(\frac{\partial\sigma}{\partial x}\right)_{t}. \tag{47}$$

We can eliminate one of the partial derivatives in equation (47) by use of the fact that along a path of constant stress equation (46) reduces to

$$0 = \left(\frac{\partial\sigma}{\partial t}\right)_{x} + U_{\sigma}\left(\frac{\partial\sigma}{\partial x}\right)_{t}, \tag{48}$$

where

$$U_{\sigma} \equiv \left(\frac{\partial x}{\partial t}\right)_{\sigma}$$

is the stress wave velocity. Thus from equations (47) and (48), we have for the change of stress along a particle path:

$$\frac{d\sigma}{dt}\bigg|_{p} = (u - U_{\sigma})\left(\frac{\partial\sigma}{\partial x}\right)_{t}. \tag{49}$$

Now if, instead of considering the stress pulse, we consider the corresponding particle velocity pulse, we can derive an equation analogous to equation (49) for the rate of change of u at a specific "particle".

$$\frac{du}{dt}\bigg|_{p} = (u - U_{u})\left(\frac{\partial u}{\partial x}\right)_{t} \tag{50}$$

where

$$U_{u} \equiv \left(\frac{\partial x}{\partial t}\right)_{u}.$$

Equations (43) and (44) can be used to eliminate the partial derivatives in equation (49) and (50) with the result along a particle path:

$$\frac{\mathrm{d}\sigma}{\mathrm{d}t}\bigg|_p = \rho(U_\sigma - u)\frac{\mathrm{d}u}{\mathrm{d}t}\bigg|_p \tag{51}$$

$$\frac{\mathrm{d}u}{\mathrm{d}t}\bigg|_p = \frac{(U_u - u)}{\rho}\frac{\mathrm{d}\rho}{\mathrm{d}t}\bigg|_p. \tag{52}$$

If equations (51) and (52) are written in differential form, manipulated and the relation $V = 1/\rho$ is used, the following integrals, valid for a specific particle path, can be derived:

$$u(\sigma) = u_0 + \int_{\sigma_0}^{\sigma} \frac{\mathrm{d}\sigma}{\rho(U_\sigma - u)} \tag{53}$$

$$V(u) = V_0 - \int_{u_0}^{u} \frac{\mathrm{d}u}{\rho(U_u - u)} \tag{54}$$

$$V(\sigma) = V_0 - \int_{\sigma_0}^{\sigma} \frac{\mathrm{d}\sigma}{\rho^2(U_\sigma - u)(U_u - u)} \tag{55}$$

$$E(\sigma) = E_0 + \int_{\sigma_0}^{\sigma} \frac{\sigma\,\mathrm{d}\sigma}{\rho^2(U_\sigma - u)(U_u - u)} \tag{56}$$

where equation (56) results from integration of equation (45), and where the zero subscripts refer to the pre-shock states of the variables. In sections 4.3 and 5.6 it will be shown that the quantities $\rho(U_\sigma - u)$ and $\rho(U_u - u)$ are measurable with in-material gauges. Thus, if suitable gauges are available to measure these quantities, the application of the equations is straightforward, requires no special assumptions about the material, and is capable of determining the entire compression and release paths necessary for the computation of propagation and attenuation. It should be re-emphasized that this is true regardless of whether the system is steady state or not. These equations must of course reduce to the Rankine–Hugoniot equations in the special case of steady flow. This case is discussed next, and the application to more general flows will be discussed in section 5.6.

4.1.2   *Steady flow: Rankine–Hugoniot equations*

In practice it requires at least a pair of stress gauges to determine $\rho(U_\sigma - u)$ and either three stress gauges or a pair of particle velocity gauges to determine $\rho(U_u - u)$ (see section 5). Thus a significant simplification will occur if $U_u = U_\sigma$. It is easy to see that this is true if particle velocity is a function of stress alone. In this case equations (53) to (56) become simpler, and a single pair of gauges of either type will be sufficient to make the measurement.

If the flow is isentropic and $U_\sigma = U_u$, then $(U_\sigma - u)$ is equal to the isentropic sound velocity C. (The subtraction of u from U_σ to obtain C is essential because U_σ is defined in laboratory coordinates, but C is defined with respect to the material.) Thus equation (53) reduces to the "Riemann" integral:

$$u(\sigma) = u_0 + \int_{\sigma_0}^{\sigma} \frac{d\sigma}{\rho C}. \tag{57}$$

Since for steady-state flow the wave profile does not change as the wave propagates, $U_\sigma = U_u = \text{constant} = U$, the "shock velocity". In addition since the profile of the steady wave is unchanging, an observer travelling with the profile will see material flowing into the shock with speed $(U - u_0)$ and flowing out with speed $U - u$. Since there can be no accumulation of mass in a steady profile, $\rho(U - u) = \rho_0(U - u_0)$. By substitution into equations (53) to (56), integration, and manipulation of the results, we obtain the usual Hugoniot relations:

$$\sigma - \sigma_0 = \rho_0(U - u_0)\,(u - u_0) \tag{58}$$

$$1 - \rho_0/\rho = (u - u_0)/(U - u_0) \tag{59}$$

$$E - E_0 = -\tfrac{1}{2}(\sigma + \sigma_0)\,(V - V_0). \tag{60}$$

Equations (58) and (59) are often combined to yield the equation:

$$\sigma - \sigma_0 = \rho_0^2(U - u_0)^2\,(V_0 - V). \tag{61}$$

These equations are valid for E, σ, u, and V either at their peak values after passage of the shock front or at any point within the shock front as the material is being compressed.

4.2 CONSEQUENCES OF THE HUGONIOT EQUATIONS

Until the relatively recent advent of sophisticated instrumentation, experimenters usually had to be content with measurements of shock and free surface velocities.[1] Such information was not adequate for

---

[1] The velocity given to an unsupported surface upon the arrival of the the shock wave at the surface. This velocity is approximately twice the particle velocity as discussed in section 5.2.

application of the analysis discussed in sections 4.1 and 5.6, and experimenters were obliged to interpret their results with the aid of the Hugoniot equations (58) to (61). Since this area has been fully reviewed (Rice *et al.*, 1958; Duvall and Fowles, 1963; Duvall, 1971), this section will contain only a brief discussion of some of the main implications of the Hugoniot equations.

It will be helpful for the later discussion to consider briefly how and why a stress wave might tend to "shock up" to a steady state condition. For simplicity we will consider that the shock is strong enough that strength effects are unimportant, and the material can be considered to be a fluid. We can write the stress-wave velocity U_σ as:

$$U_\sigma = \left(\frac{\partial\sigma}{\partial\rho}\right)^{1/2}_{\substack{\text{shock} \\ \text{path}}} + u(\sigma),$$

where

$$\left(\frac{\partial\sigma}{\partial\rho}\right)^{1/2}$$

is the sound velocity at stress σ and for the mode of compression undergone. For isentropic compression for example we would have

$$U_\sigma = \left(\frac{\partial P}{\partial\rho}\right)^{1/2}_S + u,$$

where P is the pressure and S is the specific entropy. Since for materials in most stress ranges

$$\left(\frac{\partial P}{\partial\rho}\right)^{1/2}_S$$

and $u(\sigma)$ both increase with increasing stress, such a wave would tend to sharpen up until some counterbalancing phenomenon arises to cause the wavefront to assume a steady-state form. For example, increasing heat conduction as the wave sharpened would accomplish this if no other force intervened beforehand. (In practice heat conduction can almost always be neglected and the shock and rarefaction assumed adiabatic.) For the purpose of discussion we will assume that the equilibrating force is a nonconservative strain rate dependent force $\gamma(\dot\epsilon)$ that becomes significant as $\dot\epsilon$ increases in the sharpening wavefront. Thus in this case:

$$U_\sigma = \left(\frac{\partial[P + (\gamma\dot\epsilon)]}{\partial\rho}\right)^{1/2}_{\substack{\text{shock} \\ \text{path}}} + u(\sigma).$$

In this sample, $\gamma(\dot{\epsilon})$ is 0 both at $P=0$ and at peak pressure, and the effect of "shocking up" is to increase $\gamma(\dot{\epsilon})$ between these extremes in such a way that the wavefront approaches a steady-state condition in which the term $(\partial[P+\gamma(\dot{\epsilon})]/\partial\rho)^{1/2}$ decreases with increasing pressure just enough to compensate for the increasing $u(\sigma)$. In the reverse case, in which the wave is initially too sharp, $\gamma(\dot{\epsilon})$ would be too large and the wavefront would spread out until steady state was reached.

The existence of the nonconservative $\gamma(\dot{\epsilon})$ implies an entropy change across the shock front. However, the same reasons that cause the shock to sharpen will cause the rarefraction to spread out. If it spreads out enough that $\dot{\epsilon}$ and hence $\gamma(\dot{\epsilon})$ become negligible, then the rarefaction can be considered to be an isentropic process, as is usually assumed.

4.2.1 Shock waves in fluids

As mentioned above the strength of a material can often be neglected and in early experiments usually was. According to equations (58) and (61) the variation of stress with particle velocity, or volume, is linear through the shock front of a steady-state wave. This straight line in the σ–u or σ–V planes is usually referred to as the Rayleigh line and is shown in Fig. 6 for an experiment in which the material was shocked to a peak pressure P_1. The crosshatched area under the Rayleigh line represents the internal energy of the shocked material as defined by equation (60). Also shown in Fig. 6 is the "Hugoniot", a plot of peak stresses obtained by steady shocks from σ_0, which can be obtained by a series of shock experiments. The importance of the Hugoniot is that each point on it is a well-defined thermodynamic state if the stress behind the shock is steady and free of nonconservative forces. Thus Hugoniot information, together with some theory such as the Mie–Grüneisen equation of state (see section 3), can be used to compute the P–V–E equation of state surface. Finally, the release curve, the locus of points traversed during release, is also shown. The energy recovered during release is the area under the release curve and the difference between this and the crosshatched area is the amount of energy dissipated in the material (provided $\sigma_0 \geq 0$). As discussed above, the release curve can often be considered to be an isentrope, in which case the stress will be given by equation (57). This isentrope could also be used to calculate the equation of state surface, but until recently it was not possible to measure the release curve to the precision required.

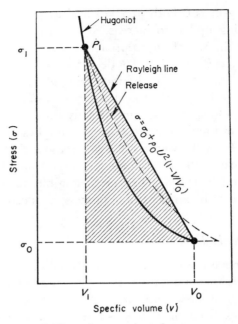

Fig. 6. Hugoniot equation of state curve.

For analysis of shock-wave propagation and interaction it is often useful to draw the Hugoniot in the σ–u plane. This is done in Fig. 7 to illustrate how a simple case, such as the collision of a flyer plate with a specimen, might be analysed. In the Figure we assume that the specimen is initially at rest and accelerated to the right upon collision.

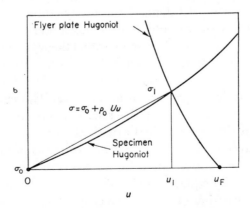

Fig. 7. Impedance match of Hugoniot curves.

(Notice that the slope of the Rayleigh line is governed by the shock impedance $\rho_0 U$.) The flyer is initially travelling at u_F and is deaccelerated to the left upon collision. Since immediately after the collision the normal components of σ and u must be the same in both materials and must lie on the Hugoniot of each material, the only possible shocked state is (σ_1, u_1). In the case discussed in section 2 for which flyer and specimen are of the same material, the Hugoniots are symmetric, and the shock particle velocity u will be exactly half the flyer velocity. Because flyer velocity is almost always easier to measure than particle velocity, flyers are often made of the same material as the specimen so that this fact can be used to determine particle velocity.

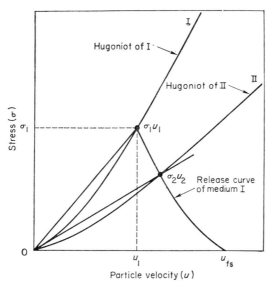

Fig. 8. Impedance match with compression and release curves.

The σ–u plane is also useful for analysing the propagation of a shock from one medium to another. To illustrate the method, consider a specific case of a shock of magnitude σ_1 travelling to the right in medium I and about to enter medium II of a lower shock impedance $\rho_0 U$. The situation is analysed in Fig. 8. Again, the boundary condition is that the normal components of u and ρ must be continuous across the interface. The new shocked state must also be on the Hugoniot of medium II and on the release curve of medium I. (The release curve of medium I slopes in the direction of higher velocities since it is being released by an acceleration of the interface into the softer material.) Thus the new state is (σ_2, u_2). It is evident that if the Hugoniot of medium II is known and either σ_2 or u_2 is measured, this is sufficient to determine a point (σ_2, u_2) on the release curve of medium I. In addition, if σ_2 and u_2 are measured in medium II and an assumption is made about the release curve, state (σ_1, u_1) on the Hugoniot of medium I can be estimated. An assumption often made for nonporous materials is that the release curve and the Hugoniot are symmetrical. In this case, if medium II is merely air or vacuum, and the free surface velocity (u_{fs} in Fig. 8) of the relieved specimen is measured, the particle velocity can be estimated to be half of the free surface velocity. Thus a measurement of σ_1 and u_{fs} is often sufficient to make a good estimate of the state (σ_1, u_1). In practice these estimates can be accurate to within several per cent for metals (Walsh and Christian, 1955). Methods for inferring the shock state in one material from measurements of another state in a second material are often referred to as impedance match methods.

4.2.2. *Effect of material strength upon shock structure*

If the material under consideration has a finite shear strength, then the stress measured by application of the Hugoniot equations is only that component normal to the wavefront, and the measurement tells us nothing of the other components. If we assume uniaxial strain and apply Hooke's law (Duvall and Fowles, 1963; Duvall, 1971), then the components of stress in the normal direction σ_x and the lateral directions σ_y and σ_z can be rewritten from equations (10) as

$$\sigma_y = (K + \tfrac{4}{3}\mu)\ \epsilon_y$$

$$\sigma_y = \sigma_z = \left(\frac{\nu}{1-\nu}\right)\ \epsilon_x.$$

The resolved shear stress is equal to

$$\tau = \tfrac{1}{2}(\sigma_x - \sigma_y) = \mu \epsilon_x = \left(\frac{1-2\nu}{2(1-\nu)}\right)\ \sigma_x.$$

Thus the shear stress increases with σ_x. This cannot continue indefinitely since every material has a shear strength, and eventually a point is reached at which the material yields. The effect of this on the Hugoniot is to cause a cusp, known as the Hugoniot elastic limit, which is shown in Fig. 9(a) by point A. Also plotted in Fig. 9(a) is the hydrostatic pressure $P \equiv \frac{1}{3}(\sigma_x + \sigma_y + \sigma_z)$. Beyond the elastic limit the "deviator stress" between P and the Hugoniot might be constant if there is no work hardening taking place, or it might be increasing or decreasing. In any case, P cannot be measured directly and must be inferred from the Hugoniot and release curves. The effect of the cusp on a shock of total strength σ_1 is shown in Fig. 9(b). In effect the section from 0 to A can be

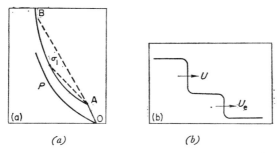

(a)   (b)

Fig. 9. (a). Hugoniot curves for a material with strength. (b) Double wave caused by cusp in Hugoniot curve.

treated as a separate shock travelling at velocity U_e and is known as the elastic precursor. Likewise the segment from A to σ_1 can be treated separately, and since its Rayleigh line (A–σ_1) has a lower slope, this segment of the wave travels at a lower velocity U. Thus the effect of the cusp is to cause a multiple shock wavefront in the stress region from A to B. For shocks of strengths above B, the velocity U would "overdrive" the elastic precursor and there would be no separation of waves.

A cusp in the Hugoniot can also be caused by a phase transition during shock compression in which the new phase has a higher density. Just as in the case described above the existence of the cusp will lead to a multiple wave structure in some stress regions. In practice, it is the observation of such a multiple wavefront that suggests the existence of possible strength or phase transformation effects. The reader is referred to Duvall and Fowles (1963), Rice et al. (1958), Duvall (1962), Zababakhin and Simonenko (1967), Cowperthwaite (1968), Lysne et al. (1971), and Johnson (1972a), for more complete discussions of this topic.

4.3   UNSTEADY WAVE SOLUTIONS

In the above discussion of Hugoniot curves it was pointed out that the Rankine–Hugoniot jump conditions hold only for steady waves or for truly discontinuous jumps in stress, density, and particle velocity. Intuitively one might expect that truly steady shock waves rarely occur, and this is in fact the case. (The conditions under which steady shock waves may occur were originally discussed by Rayleigh (1911) and Taylor (1911), and a recent review of subsequent work has been given by Kelly and Gillis (1970).) Thus, the Rankine–Hugoniot jump conditions only rarely rigorously apply.

Since present in-material gauges measure stress and particle velocity histories, we need a theoretical method to calculate these histories. In principle we need to solve only the system of differential equations formed by conservation of mass, momentum, and energy plus the constitutive relation. The behaviour of the disturbance, steady or unsteady, should then depend on the initial and boundary conditions. Since the newly developed in-material gauges follow the development of large amplitude waves, comparison of experiment and theory should be possible. Unfortunately, the general theoretical problem has never been completely solved. Nevertheless, complete analytical solutions in special cases have been obtained, and approximate numerical solutions can be obtained in most other cases. We shall next discuss these questions in more detail.

4.3.1   *Eulerian and Lagrangian coordinates*

To proceed with the discussion of unsteady wave solutions, it is necessary to understand the difference between Eulerian and Lagrangian coordinates. The Eulerian coordinate system is defined as the laboratory system, and an observer in this system sees the particles moving at velocities defined by $u = u(x, t)$. On the other hand, we may choose to describe the motion in terms of the paths of the individual particles, and the resulting system is often defined as the Lagrangian coordinate system. Let us define the positions of material particles at time t by x, where the particles are labelled by their original positions h at $t = 0$. Thus, the particle paths are defined by the function

$$x = x(h, t). \tag{62}$$

In this system, the coordinates are h and t, and the particle velocity is given by

$$u = \left(\frac{\partial x}{\partial t}\right)_h. \tag{63}$$

To an observer in this system, the particles appear stationary in the sense that the "observer" can see only the original particle positions h. Present in-material gauges actually are such Lagrangian observers, since in uniaxial strain experiments they move along particle paths and record the stress or particle velocity histories of given particles whose original positions h are known.

It can be seen that the connection between Lagrangian and Eulerian coordinates is given by the relation

$$\left(\frac{\partial}{\partial t}\right)_h = \left(\frac{\partial}{\partial t}\right)_x + u\left(\frac{\partial}{\partial x}\right)_t, \tag{64}$$

i.e. the rate of change of a quantity along a particle path is equal to the change due to its own explicit time dependence plus the "convective" change due to its change in position in laboratory coordinates. A more thorough treatment of the above topics is to be found in Courant and Friedrichs (1948).

Since $x = x(h, t)$, we can write

$$dx = \left(\frac{\partial x}{\partial h}\right)_t dh + \left(\frac{\partial x}{\partial t}\right)_h dt, \tag{65}$$

and

$$U_\sigma \equiv \left(\frac{\partial x}{\partial t}\right)_\sigma = \left(\frac{\partial x}{\partial h}\right)_t\left(\frac{\partial h}{\partial t}\right)_\sigma + \left(\frac{\partial x}{\partial t}\right)_h. \tag{66}$$

At a given constant time, conservation of mass requires that

$$\rho_0\, dh = \rho\, dx,$$

and combination of this relation with equation (65) when $dt = 0$ yields

$$\left(\frac{\partial x}{\partial h}\right)_t = \frac{\rho_0}{\rho}. \tag{67}$$

Thus, equation (66) becomes (when we recall that $(\partial x/\partial t)_h \equiv u$):

$$(U_\sigma - u) = \frac{\rho_0}{\rho}\left(\frac{\partial h}{\partial t}\right)_\sigma. \tag{68}$$

Similar reasoning shows that

$$(U_u - u) = \frac{\rho_0}{\rho}\left(\frac{\partial h}{\partial t}\right)_u. \tag{69}$$

These last two relations will be useful in the evaluation of in-material gauge records and will be discussed further in section 5.

Let us next, again, write the conservation equations for mass, momentum, and energy in uniaxial strain; first, in Eulerian coordinates we have:

$$\left(\frac{\partial \rho}{\partial t}\right)_x + u\left(\frac{\partial \rho}{\partial x}\right)_t + \rho\left(\frac{\partial u}{\partial x}\right)_t = 0$$

$$\text{(mass)} \quad (70)$$

$$\rho\left[\left(\frac{\partial u}{\partial t}\right)_x + u\left(\frac{\partial u}{\partial x}\right)_t\right] + \left(\frac{\partial \sigma}{\partial x}\right)_t = 0$$

$$\text{(momentum)} \quad (71)$$

$$\left(\frac{\partial E}{\partial t}\right)_x + u\left(\frac{\partial E}{\partial x}\right)_t - \frac{\sigma}{\rho^2}\left[\left(\frac{\partial \rho}{\partial t}\right)_x + u\left(\frac{\partial \rho}{\partial x}\right)_t\right] = 0$$

$$\text{(energy, adiabatic flow).} \quad (72)$$

Alternatively, in Lagrangian coordinates:

$$\left(\frac{\partial \rho}{\partial t}\right)_h + \frac{\rho^2}{\rho_0}\left(\frac{\partial u}{\partial h}\right)_t = 0 \quad \text{(mass)} \tag{73}$$

$$\rho_0\left(\frac{\partial u}{\partial t}\right)_h + \left(\frac{\partial \sigma}{\partial h}\right)_t = 0 \quad \text{(momentum)} \tag{74}$$

$$\left(\frac{\partial E}{\partial t}\right)_h - \frac{\sigma}{\rho^2}\left(\frac{\partial \rho}{\partial t}\right)_h = 0 \quad \text{(energy, adiabatic flow).} \tag{75}$$

Further, the engineering strain is defined by

$$\epsilon = -\frac{\partial}{\partial h}(x - h) = 1 - \frac{\partial x}{\partial h} = 1 - \frac{\rho_0}{\rho}. \tag{76}$$

In all of the above relations σ and ϵ have been defined to be positive in compression, and σ is the total stress containing thermo-mechanical and viscous components.

4.3.2   Shock-wave formation

The set of equations (70)–(72) or (73)–(75) and equation (76) gives us four equations for the five unknowns ρ, u, ϵ, σ, and E. The additional equation needed is provided by a constitutive relation specifying the dependence of σ on ρ and E, for example.

Historically, constitutive relations of the ideal gas type, or of the type (2) or (3) for relaxed fluids were used in conjunction with the above conservation equations. The resulting set of equations form a hyperbolic system and can be attacked by the method of characteristics. The standard reference for this mathematical technique applied to shock-wave problems is the book by Courant and Friedrichs (1948). Another excellent reference is the book by von Mises (1958). A good introduction to the method of characteristics is given in a short paper by Band and Anderson (1962).

When this method was applied to the problem of a piston accelerating into an initially undisturbed fluid or gas, it was found that the solution for σ, ρ, or u soon became double-valued and the method broke down. This may be termed the classic shock wave problem; it arises because a fluid or gas typically becomes less compressible with increasing compression, resulting in an increasing sound speed with increasing compression. Thus, the peak of a large amplitude disturbance travels faster than the foot, and the wave becomes increasingly steep. It is generally assumed that a "breaking" or double-valued wave is physically unreal, and the disturbance is replaced by a discontinuous shock wave at the point where the solution breaks down. These questions are discussed in detail in numerous textbooks and review articles. (See, for example, Duvall and Fowles (1963).)

The reason that the wave does not really break is usually assumed to be that viscous and thermal forces resist the "shocking-up" tendency and eventually attain a balance that determines the shock front profile. Lighthill (1956) has given a detailed exposition of this topic for the case of shock waves in gases, and Bland (1965) has discussed the related problem in solids.

Although it is felt that the physics behind the phenomenon of shock formation is fairly well understood, it has not yet been possible to solve the general shocking-up problem for arbitrary initial loading conditions and boundary constraints. Nevertheless, approximate methods have been developed that work very well in practice, and complete analytic solutions are possible for some constitutive relations in restricted regions of loading. These topics will now be pursued further.

4.3.3 *Methods of solution of unsteady wave propagation*

Integration of the set of equations discussed above is usually possible only by means of numerical finite difference techniques because the set of original equations is in general nonlinear. The two most-used methods of solution are the *method of characteristics* referred to above and the *method of artificial viscosity*. These procedures have been reviewed recently by Seaman (1970b).

In the method of characteristics, surfaces in space-time are found across which the normal derivatives of the dependent variables σ, ρ, u, ϵ, and E may have discontinuities. The original set of equations is converted to a set in which each equation contains only derivatives that are tangential to these characteristic surfaces. In one space dimension the characteristic surfaces reduce to curves in the x–t plane whose slopes correspond to wave speeds or particle velocities. The transformed equations that have only derivatives that are tangential to the characteristic surfaces are called the compatibility equations and may be numerically integrated to obtain the solution to a given problem. However, if the material has a constitutive relation that causes it to "shock-up", characteristic surfaces of the same family will intersect, resulting in a breakdown of the solution. In this case the procedure must be interrupted, and a "shock discontinuity" inserted, across which the Rankine–Hugoniot jump conditions are usually assumed to hold, although other methods are sometimes used (Seaman, 1970b).

Although the method of characteristics is often used in shock wave calculations in solids, and is very valuable for the physical insight it affords, in practice it is very difficult and time-consuming when super-sonic shock waves arise. For this reason various versions of the method of artificial viscosity, originally developed by von Neumann and Richtmeyer (1950), are often used. A review of these methods has recently been given by Seaman (1970b). In these methods the original differential equations are converted to difference equations and inte-grated numerically, but the stress is supplemented by an artificial viscous stress that is significant only in shock fronts and serves to smooth out the shock over several cells of the finite difference mesh. Artificial viscosity methods have the great advantage that they treat shock waves automatically wherever they arise, and it is not necessary to know beforehand where they will occur or to institute any special procedure when they appear. The disadvantage of such methods is that they may obscure features of the shock front that are physically important.

An objection to both the artificial viscosity methods and the charac-teristics method, in which the characteristics are artificially prevented from intersecting, is that in neither approach have we obtained a unique solution to a well-set boundary value problem. However, the overwhelming success that such methods have had in predicting and explaining experimental data makes this objection somewhat academic.

The goal of such calculations must be to predict stress and particle velocity histories along particle paths, since this information is now experimentally measurable with the in-material gauges to be described in the next section. In the remainder of this section we shall review the results of applying the above numerical methods of solution to wave propagation problems in materials that are elastic–plastic, elastic–viscoplastic, and stress relaxing. Wave propagation in porous materials and composites will also be discussed.

4.3.4   Wave propagation in rate-independent, elastic–plastic solids

In this case the appropriate constitutive relations are given by equations (10) or (11), depending on whether the solid has been stressed beyond its yield point. Restricting ourselves to the direction of wave propagation in an infinitesimal uniaxial strain condition, we rewrite equation (10) as the constitutive relation for stresses below the yield point:

$$\sigma = (K + \tfrac{4}{3}\mu)\ \epsilon. \tag{77}$$

From equation (76) it is seen that

$$\frac{\partial \epsilon}{\partial t} = -\frac{\partial u}{\partial h}. \tag{78}$$

Taking the time derivative of equation (77) and combining it with equation (78) yields

$$-\frac{\partial u}{\partial h} = \frac{1}{(K + \tfrac{4}{3}\mu)}\left(\frac{\partial \sigma}{\partial t}\right), \tag{79}$$

where we assume K and μ are constants in the elastic region. Operation on equation (79) with $\partial/\partial h$ yields

$$-\frac{\partial^2 u}{\partial h^2} = \frac{1}{(K + \tfrac{4}{3}\mu)}\frac{\partial}{\partial t}\left(\frac{\partial \sigma}{\partial h}\right). \tag{80}$$

Combination of equation (80) with (74) yields

$$\frac{\partial^2 u}{\partial h^2} - \frac{1}{C_l^2}\left(\frac{\partial^2 u}{\partial t^2}\right) = 0, \tag{81}$$

where $C_l = \{(K + \frac{4}{3}\mu)/\rho_0\}^{1/2}$, the longitudinal elastic sound speed. Thus, as we expect, in the elastic region we obtain the well-known elastic wave equation for the particle velocity.

This result suggests that the set of equations (73–75), (76), and (77) is hyperbolic and lends itself to analytic solution by the method of characteristics. This is in fact the case.

In the plastic region, the equation (77) is replaced by one of the relations from equation (11):

$$\sigma = P(\epsilon) + \tfrac{2}{3}Y \tag{82}$$

or

$$\frac{\partial \sigma}{\partial t} = \frac{\mathrm{d}P}{\mathrm{d}\epsilon}\left(\frac{\partial \epsilon}{\partial t}\right) = -\frac{\mathrm{d}P}{\mathrm{d}\epsilon}\left(\frac{\partial u}{\partial h}\right) \tag{83}$$

$$= -K\left(\frac{\partial u}{\partial h}\right),$$

where we have assumed a constant yield point and have used equation (11). We next operate on equation (83) with $\partial/\partial h$ to obtain

$$\frac{\partial}{\partial t}\left(\frac{\partial \sigma}{\partial h}\right) = -K\left(\frac{\partial^2 u}{\partial h^2}\right) - \frac{\partial K}{\partial h}\left(\frac{\partial u}{\partial h}\right). \tag{84}$$

Combining equations (84) and (74) we obtain

$$\frac{\partial^2 u}{\partial h^2} - \frac{1}{C^2}\left(\frac{\partial^2 u}{\partial t^2}\right) + \frac{\partial u}{\partial h}\left(\frac{\partial lnK}{\partial h}\right) = 0, \tag{85}$$

where C is the bulk sound speed $(K/\rho_0)^{1/2}$. Equation (85) is a damped wave equation and indicates that in the region above the yield point small disturbances travel with the bulk sound speed $(K/\rho_0)^{1/2}$. Since the bulk modulus K increases with compression for most materials, the shocking-up referred to earlier occurs in the plastic region, and an analytic solution is not attainable. Thus, one of the approximation methods discussed above must be used. An illustration of the power of the artificial viscosity method is given in Fig. 10, which shows calculated wave profiles in 2024-T8 aluminium. As can be seen, the effect of the elastic–plastic transition is to cause an elastic precursor wave to run out ahead of the main plastic shock wave. At sufficiently higher peak stresses the plastic shock wave would travel at a velocity faster than the elastic velocity, and only one wave would be present (Duvall and Fowles, 1963).

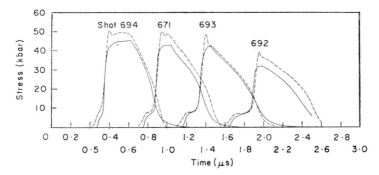

Fig. 10. Comparison of computed and measured stress histories from impacts with 2024-T8 aluminium at 50 kbar. ——— Transducer stress. – – – Stress calculated from puff code. (Histories correspond to depths in target of 3.15 mm, 6.26 mm, 8.90 mm, and 12.05 mm from the impact plane.) (From Seaman *et al.*, 1969.)

When unloading of an elastic–plastic material occurs after initial loading, it is often observed that the unloading or release path is not symmetric to the loading path, as shown in Fig. 3, but behaves as though the yield strength and shear modulus changed continuously from the point at which unloading begins. This effect, called the Bauschinger effect, can be arbitrarily modelled by making the shear modulus and the yield strength functions of stress, strain, and loading history. Such a procedure is discussed by Seaman (1970c), and an example of such a nonsymmetrical loading–unloading path generated by this procedure

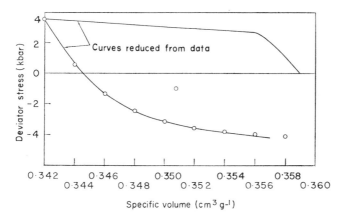

Fig. 11. Relationship between deviatoric stress and specific volume obtained from 50-kbar tests in 2024-T8 aluminium. ○, Unloading points calculated. (From Seaman *et al.*, 1969.)

is shown in Fig. 11. The increase in shear modulus upon unloading causes the attenuation of peak stress to occur significantly faster than would be the case if the modulus remained constant. (The Bauschinger effect was included in the computed stress histories shown in Fig. 10.) Herrmann *et al.* (1970a) have discussed an alternate numerical approach to the Bauschinger effect based on a model originally developed by Duwez (1935). In this model, a material particle is considered to consist of a number of elastic, perfectly plastic elements connected in parallel, where each element has a different constant yield stress. As the material is loaded and unloaded, the elements yield one at a time to produce a resultant loading–unloading–reloading curve with hysteresis.

4.3.5 *Wave propagation in elastic, viscoplastic solids*

Consider a material that is described by the constitutive relation (17). Operating on this equation by $\partial/\partial h$ yields

$$\frac{\partial \sigma}{\partial h} = \frac{\partial}{\partial h} (P + \tfrac{2}{3} \Upsilon) - \tfrac{4}{3} \eta \left(\frac{\partial^2 u}{\partial h^2} \right)$$

where we have used equation (78). Combination with equation (74) yields

$$\frac{\partial u}{\partial t} = \frac{4\eta}{3\rho_0} \left(\frac{\partial^2 u}{\partial h^2} \right) - \frac{1}{\rho_0} \left(\frac{\partial}{\partial h} \right) (P + \tfrac{2}{3} \Upsilon). \tag{86}$$

This is a diffusion equation and suggests that the set of conservation equations plus the constitutive relations form a parabolic set. This is in fact the case. A property of diffusion equations is that their solutions are physically unreasonable at small times, since they are associated with infinite propagation velocities (Band, 1959). This is a feature of parabolic equations in general, and, as discussed earlier, makes such constitutive relations impractical for transient problems.

4.3.6 *Wave propagation in stress-relaxing solids*

The type of constitutive relation currently receiving the most attention in shock-wave work is that for the stress-relaxing solid, as given by equation (23). A special case is the relation for a Maxwell solid, equation (25). If we operate on equation (25) by $\partial/\partial h$, we obtain

$$\frac{\partial \sigma}{\partial h} = \frac{\partial}{\partial h} (P + \tfrac{2}{3} \Upsilon) - (K + \tfrac{4}{3} \mu) \, T \left(\frac{\partial^2 u}{\partial h^2} \right) + \rho_0 \, T \left(\frac{\partial^2 u}{\partial t^2} \right), \tag{87}$$

where K and μ are here assumed to be constants, and where we have used equation (78). Again if this is combined with equation (74) we obtain

$$\rho_0 \frac{\partial u}{\partial t} = -\frac{\partial}{\partial h} \left(P + \tfrac{2}{3}\varUpsilon\right) + \left(K + \tfrac{4}{3}\mu\right) T\left[\frac{\partial^2 u}{\partial h^2} - \frac{1}{C_l^2}\left(\frac{\partial^2 u}{\partial t^2}\right)\right], \qquad (88)$$

where C_l is again the elastic longitudinal sound speed:

$$C_l = \left(\frac{K + \tfrac{4}{3}\mu}{\rho_0}\right)^{1/2}.$$

This is a damped wave equation (Malvern, 1951) and suggests that the complete set of governing equations is hyperbolic with characteristic speeds equal to $\pm C_1$. This is in fact the case (with an additional characteristic direction being along the particle paths), and various forms of equation (23) have been used with success to describe wave propagation in the stress regime where elastic precursor waves are observed. A detailed review of wave propagation in materials of this type has recently been given by Herrmann et al. (1970b).

In a plate impact experiment the initial loading may be considered to be almost instantaneous, the rate terms in equation (23) dominate, and the instantaneous stress–strain curve is given by

$$\sigma = \left(K + \tfrac{4}{3}\mu\right) \epsilon. \qquad (89)$$

If this relation is linear (K and μ constant) the analysis is simplified considerably, and the rate of decay of the elastic precursor can be easily obtained in the manner of Taylor (1965) and succeeding workers. However, as pointed out by Herrmann et al. (1969, 1970b), there is increasing evidence that the instantaneous response curve is nonlinear (i.e. K and μ in equation (88) are functions of strain) and that the shape and decay of the elastic precursor is a strong function of this nonlinearity.

In the paper by Herrmann et al. (1970b) the governing equations were solved by a characteristic method for the linear instantaneous response curve and by an artificial viscosity method for the nonlinear case. They used a dislocation model for the $G(\sigma, \epsilon, t)$ of equation (23) and found that dislocation multiplication plus the nonlinearity of the instantaneous response caused a sharp spike in the elastic precursor wave that is probably not resolvable by present experimental stress or particle velocity gauges. It thus appears that numerical solutions for wave propagation in stress-relaxing solids are currently more detailed than can be experimentally measured; better experimental resolution of stress and particle velocity histories is needed to help determine the best constitutive relations for a given solid.

The problem is compounded by the current ambiguity in dynamic dislocation models. As discussed earlier, there is a variety of possible dislocation density and mobility functions to choose from when one attempts to predict transient shock behaviour in crystalline solids. Possibly the most unambiguous results recently reported are those by J. N. Johnson et al. (1970), in which theoretical elastic precursor decay in uniaxial strain conditions for single crystals of Cu, W, NaCl, and LiF was calculated from the best available dislocation models for several crystal orientations. The theoretical predictions were compared with experimental results from plate-slap experiments. Although the theory predicted the proper relative order of the precursor amplitudes in different propagation directions, unrealistically high mobile dislocation densities were required to force agreement with the experimental data. Although it is possible that the experimental uncertainties and the nonlinearity of the instantaneous response curve referred to above may resolve the discrepancy, it must be concluded that good understanding of the transient response is not yet attained. Nevertheless, there is little doubt that such dislocation models, combined with an understanding of the instantaneous response curve, will eventually explain the response of crystalline solids to dynamic loading.

4.3.7   Wave propagation in viscoelastic materials

Shock waves in polymers typically show a steep initial rise followed by a more gradual approach to a final stress state. Such a stress record in C-7 epoxy is shown in Fig. 12. Schuler (1970) and Barker and Hollenbach (1970) have found that in PMMA for stresses less than 6 kbar this type of behaviour can be explained as that of a simple viscoelastic material. (For higher stresses the above authors conclude that plastic flow occurs.) In such materials a nonlinear instantaneous elastic response causes the material to shock-up, but this tendency is opposed by a stress relaxation effect.

Schuler found that the waves produced in PMMA were steady, and were of a particular type allowable under general viscoelastic theory, namely steady supersonic waves that exhibit a discontinuous jump in particle velocity followed by a slow rise to the maximum value attained.

The shape of the slowly rising part of the wave is determined by competition between the shocking up tendency of the material and the stress relaxation. Schuler showed that suitable choices of material parameters in this model could produce agreement of calculated wave profiles with the experimental data for PMMA.

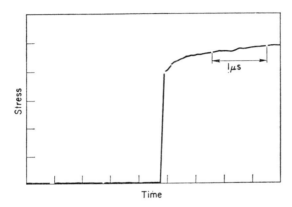

Fig. 12. Stress–time profile from a manganin gauge embedded in C–7 epoxy. Note the sharp rise in stress followed by a more gradual rise.

It should be expected that stress waves in the low kilobar range in other polymers may also be described by this type of viscoelastic model.

4.3.8 Wave propagation in porous solids and composites

As discussed earlier, a variety of constitutive relations have been developed for porous solids and composites. Since they often are based on a macroscopic approach that treats the porous solid or composite as an equivalent homogeneous material, the methods of solving the wave propagation are basically the same as described above. Most of the calculations have used artificial viscosity in combination with a rate-independent, elastic–plastic, constitutive relation, although, as mentioned earlier, Johnson (1968) and Butcher (1970) have used a stress-relaxation model in porous materials.

Figure 13 shows computed and experimental stress histories in a porous iron target reported by Seaman (1970a); the constitutive relation was of the type used by Seaman and Linde (1969) and is illustrated in Fig. 5.

A porous solid is a special type of two-component composite, with one component being void. If the method of mixtures discussed earlier is used to derive an average constitutive relation for a composite (similar to the way that such constitutive relations are derived for porous materials), then the standard computation techniques can be used to calculate wave propagation in composites. However, this assumes that the scale of the inhomogeneities is small compared with the scale of

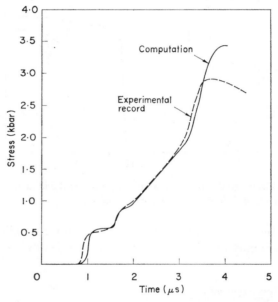

Fig. 13. Record of manganin gauge embedded in epoxy behind a porous iron target that was impacted by a thin solid iron flyer. (From Seaman, 1970a.)

measurement and interest. When this is not the case, the calculation must handle each component separately, as was done by Lundergan and Drumheller (1970) in a PMMA–steel laminate. Another important question is whether the material behaves in a linear or nonlinear elastic manner or whether one or more of the composite components exhibits plastic flow or porosity. The effect of nonlinear elasticity on wave propagation is shown in Figs 14a and 14b, which are taken from the paper by Lundergan and Drumheller (1970). It can be seen that the nonlinearity of one of the composite components had a large effect on the stress history. The effects of plasticity or porosity may also be expected to be large.

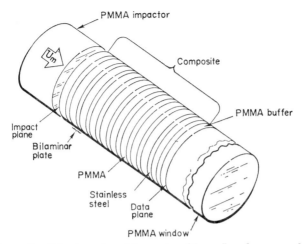

Fig. 14a. Schematic of the target and impactor. (From Lundergan and Drumheller, 1970.)

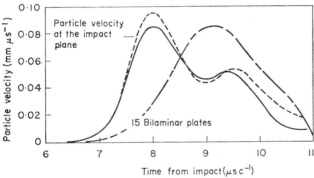

Fig. 14b. Comparison of the measured transmitted waveforms with the calculations of the wave propagation codes. The artificial viscosity code accounts for nonlinear behaviour of the PMMA, whereas the characteristic code treats both materials as linear-elastic. The experimental data were obtained with a laser interferometer. ———Data. – – – Artificial viscosity code. –·– Characteristic code. (From Lundergan and Drumheller, 1970.)

4.4 CURRENT PROBLEMS IN WAVE PROPAGATION CALCULATIONS

We summarize the above discussion by listing the areas of current interest and activity in calculating stress histories in solids.

a. *Effect of nonlinearity of the instantaneous response curve.* As discussed above, recent work indicates that the initial elastic behaviour of solids is a strong function of the nature of the instantaneous stress–

strain curve. That is, the response of the material before any relaxation process (such as dislocation motion) has time to occur has a large effect on subsequent material response. Thus, the nature of the instantaneous response curve is currently receiving attention.

b. *Finite strain models.* Methods of characterizing finite strain, and the effect of finite strain on wave propagation, are currently being studied by many investigators.

c. *Dislocation models.* The most important relaxation mechanisms in crystalline solids for determining wave propagation behaviour are dislocation density, multiplication, and motion. Various dislocation multiplication, and mobility functions are currently being proposed and the resulting predicted wave propagation behaviour is being compared with experiment. As yet, however, insufficient knowledge of dislocation behaviour is available to make unambiguous predictions possible.

d. *Constitutive relations during unloading and in tension.* The attenuation of a shock wave is determined by the unloading path taken by the material. It is difficult to measure the constitutive relation under dynamic conditions when the material is unloading into tension. Since the unloading behaviour is so important in determining attenuation, it is important to develop methods for measuring the material response in this regime.

e. *Wave propagation in porous solids and composites.* These materials pose special problems for wave propagation calculations, and much current effort is being spent to characterize their properties properly.

5   Instrumentation and measurement techniques

The earliest use of shock waves was to obtain equation of state data in a region of pressure and temperature not accessible by any other means. If it is assumed that the Hugoniot equations (equations (58) and (59)) apply, then a measurement of any two of the variables σ, u, U, or $V(V=1/\rho)$ will allow the other two to be calculated. Much of the experimental effort of workers in the field of shock waves has been directed towards measuring the above parameters and determining the Hugoniot equations of state of many solids and liquids.

The shock velocity U is the most easily and directly measured of the above four variables. Until recently, the particle velocity u has been inferred from measurements of the velocity given a free surface by the shock wave. Transducers have been developed that measure the component of the stress in the direction of wave propagation, but attempts to measure density or volume directly have met with only limited success.

Within the last ten years, interest in the application of shock waves has broadened to include studies of other material properties, e.g. magnetic properties, electrical conductivity and polarization, optical properties, and mechanical properties. Many techniques have been developed to make measurements that would assist in the study of these properties.

Experimental techniques have been reviewed by Deal (1962), Doran (1963), Duvall and Fowles (1963), Rice *et al.* (1958), and Keeler and Royce (1971, pp. 51–80, 138–150). In this review, we will discuss in detail only those methods that have been developed since these reviews were published. However, for completeness and to show the historical development of the experimental techniques, we will mention all methods that have achieved wide usage. For convenience, the experimental techniques discussed below are grouped according to the particular parameter being measured. Following the discussion of specific experimental methods, we will describe the use of multiple-embedded gauges and the analysis of the data obtained with them.

5.1 MEASUREMENT OF STRESS

Stress or pressure transducers must possess some property that is stress or pressure dependent in a measurable way. Such transducers are secondary measuring devices, and their calibration is ultimately based on measurements of shock and particle velocity that can be measured directly. These transducers may be grouped according to the nature of this stress-dependent property, and in the discussion that follows this will be done.

5.1.1 *Piezoelectric transducers*

Probably the first use of a piezoelectric material to measure shock stresses was by Goranson *et al.* (1955). These authors used tourmaline and reported measurements up to 324 kbar in explosively-loaded iron, including an elastic wave of 15.7 kbar. Hearst *et al.* (1964, 1965) reported on the piezoelectric response of tourmaline for shock strengths up to 21 kbar. Although tourmaline was the first piezoelectric material to be used in shock work, it has been displaced by quartz, perhaps because the advent of synthetic quartz made it possible to obtain flaw-free material with a minimum of deviation between samples.

The Sandia quartz gauge was developed in the early 1960's, and the principles of its operation have been discussed in several papers. We refer the reader to Jones *et al.* (1962), Neilson *et al.* (1962), Halpin *et al.*

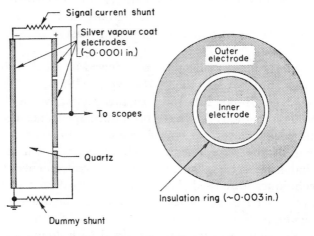

Fig. 15. Guard-ring electrode configuration. The gauge output current is measured across a small ($\sim 10\ \Omega$) resistive shunt. The voltage on the outer electrode is balanced by a shunt resistor. (After Halpin *et al.*, 1963.)

(1963), Graham *et al.* (1965a), and Ingram and Graham (1970). The gauge is a disk of synthetic X-cut quartz with a large diameter-to-thickness ratio to ensure one-dimensional strain in the gauge.

In Fig. 15, we show the basic features of a guard-ring Sandia gauge. For proper use the gauge must be oriented in a positive manner, since a dielectric anomaly has been observed in negatively oriented quartz crystals (Graham, 1962). The gauge is said to be positively oriented if a compressive stress wave propagates from the $-X$ to the $+X$ electrode.

The Sandia gauge has been used in three basic configurations: (1) fully electroded without a guard ring (Jones *et al.*, 1962); (2) with a guard ring (Graham *et al.*, 1965a); (3) with a shorted guard ring (Jones and Halpin, 1968). To use the fully electroded quartz gauge, a large diameter-to-thickness ratio (~ 5) is required to ensure that one-dimensional strain occurs during the stress profile measurement. A 5-to-1 ratio, assuming a $1\frac{1}{2}$-in. diameter gauge, implies a thickness of $\frac{1}{4}$ in. and thus a recording time of about 1 μs. The use of the guard-ring geometry, as shown in Fig. 15, allows one to use a thicker gauge, for a given diameter, and hence to increase the effective recording time. With the shorted guard-ring geometry, the dummy shunt is removed, and the outer electrode is shorted to the input electrode by coating the gauge periphery with a conducting material. Jones and Halpin (1968) claim that there is no evidence of differences in gauge characteristics between the guard-ring and shorted guard-ring geometries.

There is a fundamental difference in the mode of operation between the Sandia quartz gauge used in shock-wave measurements and more conventional piezoelectric transducers. The Sandia gauge operates in short-circuit mode, i.e. the short-circuit piezoelectric current is proportional to the specimen–quartz interface stress for times less than the wave transit time through the quartz. The more conventional mode of transducer operation is to record the piezoelectric charge output on a time scale that is long compared with the stress wave transit time in the gauge. In this case, the entire gauge is assumed to reach stress equilibrium, and the charge output is proportional to the average stress in the gauge. This type of gauge provides recording times that are dependent only on the rate of loss of charge, and therefore the gauge records for extremely long times whereas the Sandia gauge records for only one wave transit, of the order of a few microseconds. However, the outstanding characteristic of the Sandia gauge is the excellent time resolution that can be achieved. The time resolution of the gauge is limited only by the frequency response of the recording instrumentation and by any shock tilt that might be present.

The short-circuit piezoelectric current can be related to the instantaneous stress at the quartz–specimen interface using electrostatic theory (Jones et al., 1962; Neilson et al., 1962; Graham et al., 1965a). For times less than the wave transit time in the gauge and for stresses less than 25 kbar, the result is

$$i(t) = \frac{Akv}{h} \left[\sigma_{xq}(0, t) - \sigma_{xq}(h, t) \right] \tag{90}$$

where x is the distance measured from the quartz–specimen interface in the direction of wave propagation, v is the elastic wave velocity in the quartz, t is the time, A and h are the area and thickness of the quartz gauge, σ_{xq} is the x component of the stress in the quartz for conditions of one-dimensional strain, and k is the one-dimensional strain piezo-electric coefficient. Note that for $t < h/v$, $\sigma_{xq}(h, t)$ is zero and

$$\sigma_{xq}(0, t) = \frac{hi(t)}{Akv}. \tag{91}$$

For $t \geq h/v$, the current is proportional to the difference in stress at the two gauge surfaces. The value of k is given by Ingram and Graham (1970) as

$$k = [2.011 \times 10^{-8} + 1.07 \times 10^{-10}\ \sigma]\ \text{C cm}^{-2}\ \text{kbar}^{-1}.$$

The Sandia gauge has been calibrated by impacting quartz gauges with quartz-headed projectiles. Then the relation

$$\sigma = \rho_0 U u \tag{92}$$

can be used to compute σ since the particle velocity u is one half the impact velocity and the shock velocity U is also measured in the same experiment. For calibration details see Graham $et\ al.$ (1965a) and Ingram and Graham (1970).

Equations (90) and (91) are based on a linear theory; exceptions to this theory are discussed by Neilson and Benedick (1960). Graham $et\ al.$ (1965a), Graham and Halpin (1968), Graham and Ingram (1968, 1969), and Freeman (1969), Canfield and Cutchen (1966) discussed the effect of ionized impurities. Despite these exceptions (many of which are negligible), a positively oriented X-cut quartz gauge, used properly, has an accuracy of ± 2.5 per cent up to 25 kbar (Graham $et\ al.$, 1965a; Ingram and Graham, 1970). Current work at Sandia Laboratory indicates that the accuracy of this gauge can be improved and perhaps the stress range may be extended (Graham and Ingram, 1969).

The quartz gauge may be used in explosive loading experiments (e.g. Jones and Mote, 1969) or in impact experiments with a light-gas gun. The quartz gauge may be used in several ways in impact experiments: (1) to measure impact stresses propagated through a target; (2) as the impacting head of a projectile; (3) as the target to be impacted; (4) a combination of these geometries; and (5) as a reverberation plate in successive loading and unloading.

In case (1) the gauge is placed on the sample opposite the impact face, and the stress–time profile at the sample gauge interface is determined (e.g. Graham *et al.*, 1965a; Linde and Schmidt, 1966b; Jones and Mote, 1969). In Fig. 16, the stress–time profile obtained from a gauge in direct contact with [100] oriented single crystal sodium chloride is shown. The profile is that at the quartz–specimen interface. The stress in the sample is obtained from the measured stress in the quartz by the impedance match method (see section 4.2.1).

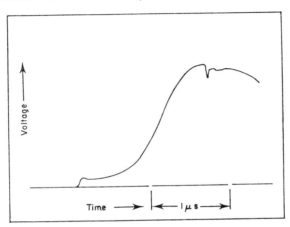

Fig. 16. Voltage–time profile obtained from a quartz gauge in direct contact with an impact-loaded single crystal of sodium chloride. Note the elastic precursor.

In case (2) the quartz gauge is made the head of the projectile and impacts the sample directly (e.g. Halpin *et al.*, 1963; Rohde, 1969a). This impact geometry has the advantage that the stress measured in the quartz is also the stress in the sample because of the boundary condition that the stress be continuous across the boundary between the two impacting materials. In addition, this geometry allows the target to be heated or cooled without interfering with the quartz gauge (e.g. Rohde, 1969a). The major disadvantage is the problem of making electrical contact with the gauge mounted in the projectile.

Normally the quartz gauge is limited to stresses below about 25 kbar. However, this limit can be extended by using the quartz gauge as a projectile head and facing the gauge with a high-impedance material, e.g. sapphire or tungsten. Because of the impedance mismatch between the quartz and the higher-impedance material, the impact stress in the target (and in the gauge facing) will be greater than that in the quartz gauge. The Hugoniot of quartz and the high-impedance material are then used to obtain the stress in the target material by the impedance match technique (see Ingram and Graham, 1970).

In case (3), the material to be studied is used as the projectile head and impacts a quartz gauge used as a target (e.g. Halpin *et al.*, 1963; Halpin and Graham, 1965). Using a projectile head of X-cut quartz, several investigators have used this geometry to study the piezoelectric response of quartz (e.g. Graham, 1961; Graham *et al.*, 1965a; Graham and Ingram, 1968, 1969).

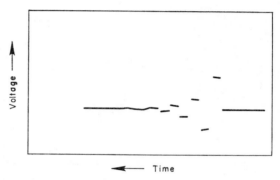

Fig. 17. Output from a quartz gauge used as a reverberation disk between two specimens of epoxy. The direction of wave propagation reverses at each transit in the quartz; hence successive reverberations cause the polarity of the output current to alternate. The amplitude of the output steps is proportional to the stress difference between the two gauge faces. Note that this stress difference decreases with successive reverberations and approaches zero as the epoxy–quartz–epoxy system comes to equilibrium. (After Lysne *et al.*, 1969.)

In cases (4) and (5), combinations of the above mentioned geometries are used. Two independent stress–time profiles are obtained when a quartz gauge is used both as an impactor and as a target backing, i.e. a combination of cases (2) and (1) (Halpin *et al.*, 1963). Halpin and Graham (1965) used this combination to compare the impact and propagated stress profile in Plexiglas. Lysne *et al.* (1969), in addition to using quartz gauges in geometries (1) and (2), also used them as a reverberation disk between the impacting projectile and the target. Depending on the impedance of the reverberation disk relative to the other two materials, each reverberation successively loads or unloads the projectile-head material. In this manner, points on a release adiabat or on recentered Hugoniot curves can be obtained for the projectile-head material. These workers showed that the quartz gauge can be successively loaded and unloaded several times. Their results for an epoxy resin projectile impacting a quartz gauge reverberation disk backed by epoxy resin is shown in Fig. 17. The gauge output is proportional to the difference in stress between the two gauge faces, and after the initial shock transit in the quartz, the stress is non-zero on both faces. A correction factor may be necessary when the reverberation technique is used (see Lysne, 1972).

Rohde and Jones (1968) and Towne (1970) made measurements and developed techniques so that quartz gauges may be used on heated specimens up to 573 K. Jones (1967) was able to extend the useful temperature of the quartz gauge to 79 K. Some gauge packaging techniques are discussed by Jones *et al.* (1962), O'Brien and Wasley (1966), and Ingram and Graham (1970). The paper by Ingram and Graham (1970) is the most comprehensive one on the quartz gauge, its use, and special construction and assembly techniques.

Recently quartz gauges have been used to record the stress pulses produced in thin targets as a result of laser illumination. In this type of experiment, the duration of the stress pulse is generally less than the wave transit time in the gauge. Graham and Ingram (1972) have discussed the use of the quartz gauge for these measurements.

The main advantages of the quartz gauge are its excellent time resolution and its simplicity. The main disadvantages are the upper limit in stress, the restriction on the recording time of the gauge, and that the stress–time profile measured is that in the quartz. Except for the case in which the gauge is the projectile head, the experimenter is required to use the impedance match method to obtain the stress–time profile in the specimen material. Jones *et al.* (1962) discussed the problem of obtaining the sample stress–time profile from the quartz stress–time profile when stress relaxation and yielding are present.

Graham (1972a, 1973) has shown that lithium niobate may also be used as a stress gauge. The piezoelectric stress constant of lithium niobate is an order of magnitude larger than for quartz. Thus a gauge from this material would be useful for low-amplitude stress measurements or where gauge miniaturization is desirable.

5.1.2   *Piezoresistive transducers*

For many years it has been known that the electrical resistance of materials, conductors, semiconductors, and insulators was a function of

4

the pressure or stress applied to the material, i.e. that materials in general are piezoresistive. A large volume of data from static experiments is available (Bridgman, 1952; Balchan and Drickamer, 1961a; Bundy and Strong, 1962; Samara and Drickamer, 1962). Only relatively recently have resistance measurements on shock-loaded materials been made. Alder and Christian (1956), measured the electrical conductivity of ionic and molecular crystals of CsI, I_2, CsBr, and LiAlH$_4$ up to shock stresses of about 250 kbar.

Hauver (1960) and Eichelberger and Hauver (1962) seem to have been the first to exploit the piezoresistive nature of materials to measure pressure in a shock-loaded sample. They used a thin disk of sulphur to measure the stress–time profile in detonating Baratol explosive. Berg (1963) further studied the use of sulphur as a stress transducer and showed that the sulphur would respond to a second stress increase. The principal disadvantage of sulphur is that it is useful only at stresses above about 70 kbar. At this stress, there is a dramatic decrease in the resistivity, and below 70 kbar the resistivity is too large for a useful piezoresistive device. The first shock work with metals was by Bernstein and Keough (1962, 1964) and Fuller and Price (1962) who measured the electrical conductivity in thin wires embedded in epoxy. Bridgman (1950) used manganin as a secondary pressure gauge and reported a linear resistance change with static pressure up to 30 kbar. Fuller and Price (1962) reported that this linearity with pressure was valid for dynamic stresses up to 300 kbar and suggested that manganin would be a useful material for a dynamic stress transducer.

Manganin is particularly suited for dynamic measurements because its temperature coefficient of resistance is almost zero even at temperatures near melting (Keough, 1964); this allows first-order separation of pressure and temperature effects. However, sufficient data are not available to know how the temperature coefficient of resistance may depend on pressure. Keough et al. (1964) showed that the pressure coefficient of resistance for manganin is independent of temperature between 25 and 500°C up to at least 150 kbar.

The ability of fine manganin wires or foils to respond faithfully to stress–time profiles has been shown by several workers: for example, a two-wave structure in glass and granite between 100 and 200 kbar (Fuller and Price, 1964), multiple waves from a shock reverberating between materials of different impedance (Fuller and Price, 1964; Keough et al., 1964), successive impact of two flying plates and detection of lateral spalling of an explosively accelerated flyer plate (Fuller and Price, 1964). Fuller and Price (1965, 1969) and Kusubov and van Thiel (1969) used manganin transducers to study the shock and release waves in magnesium and aluminium.

The study of release waves using manganin is complicated by piezoresistive hysteresis in the transducer, i.e. the releaes path in the stress-resistance plane does not coincide with the locus of loading points used as the loading path calibration. Keough *et al.* (1964) measured the residual changes in the transducer resistance at zero stress by placing an air cavity directly behind the transducer wire. In this geometry, the transducer is stress loaded and then relieved to zero stress, and any resistance change can be readily detected. Their results indicate that the residual resistance change is a function of peak stress and is about 7 per cent of the original resistance at 150 kbar. The exact cause of this hysteresis is at present not known, but some possible causes will be discussed later in connection with transducer calibration and use.

Several different transducer geometries have been used. Figure 18 shows four transducer designs which are of two types: (1) two-terminal and (2) four-terminal. The advantages and use of each type will be discussed later in connection with the use of piezoresistive transducers. The manganin portion of the transducer may be made from thin wire (usually 0.001 to 0.003 in. in diameter) or be photoetched from manganin foil (usually 0.0005 to 0.002 in. thick). The resistance of the transducer may be varied by changing the length, thickness, or diameter of the transducer element. Generally resistances of 10 to 50 Ω are used to record shock amplitudes in the low kilobar region (less than about 20 kbar) because of the increase in transducer sensitivity. At the higher stress amplitudes transducer resistances are about 0.5 to 2 Ω.

The power supply for the manganin transducer can be of the constant current or bridge type. In Fig. 19, the schematic for the constant current pulsed-power supply developed at Stanford Research Institute (SRI) is shown. This design has a capacitor as the essential element, although an inductor may be used in place of the capacitor (see Grady, 1969). The power supply provides a triggered, constant, viewing current to the transducer and is pulsed to prevent gauge destruction. The constant current is produced by capacitor discharge through a series of ballast resistors.

In the circuit of Fig. 19, the ballast resistance is composed of R_6, the crowbar source resistor, and the 50-Ω resistors in each of the power supply output legs, which are also used for cable terminations. Excessive wire heating caused by the dissipation of power in the transducer is controlled by limiting the duration of the current pulse. The pulse length can be adjusted by using the "crowbar adjust" potentiometer (see Fig. 19); for the circuit values shown, the pulse length is adjustable

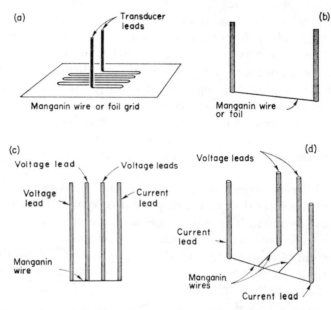

Fig. 18. Manganin transducer designs showing two- and four-terminal geometries. (a) Two-terminal wire or foil grid geometry. (b) Two-terminal straight wire or foil geometry. (c) Four-terminal straight wire or foil geometry. (d) Four-terminal wire or foil in π geometry.

between about 20 and 300 μs. During the experimental setup, the transducer is usually pulsed several times, and the ability to "crowbar" (discharge) the power supply lessens the chance of destroying the transducer.

The simplest form of recording system for the four-terminal transducer is shown in Fig. 20 (Keough, 1968). The signal generated by the change in resistance of the transducer element is recorded with oscilloscopes. An idealized waveform showing the usual sequence of events is presented in Fig. 21. Generally an oscilloscope is used to record the overall current turn-on step and the stress pulse; this allows one to

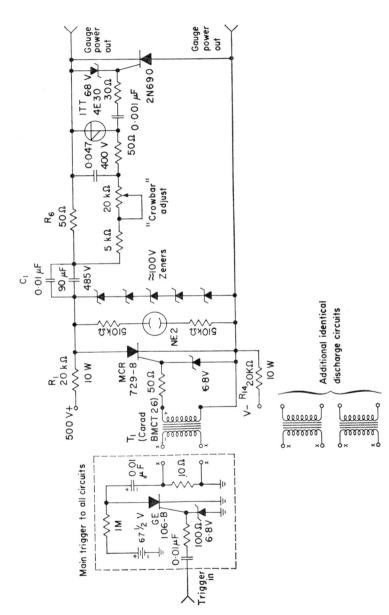

Fig. 19. Constant-current power supply with diverting (crowbar circuit).

obtain the ratio $\Delta V/V$, where V is the voltage across the gauge and ΔV is the change caused by the stress wave (see Fig. 21). Additional oscilloscopes may be used at faster sweep speeds and higher sensitivities to record merely the stress pulse and thereby record the details of the stress–time profile.

When using a constant current system, it is important that the change in transducer resistance due to stress be very small compared with the total ballast resistance. If this condition does not hold, the current through the transducer will not be constant, and the reduction of the transducer data will be complicated (see Keough, 1968). Generally a non-constant current situation arises when the transducer resistance is

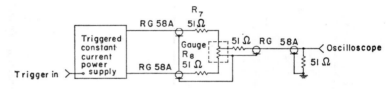

Fig. 20. Four-terminal manganin recording circuit (schematic).

in excess of 2 Ω (for a power supply with circuit parameters, as shown in Fig. 19). It is possible to increase the ballast resistance but with a resultant reduction in current and hence a loss of transducer sensitivity.

A related problem, encountered when the transducer resistance is increased, is that of shunting by the transducer viewing circuit. In an ideal situation, no current would flow in the circuit used to measure the voltage produced in the transducer (see Fig. 20). In practice, however, some current does flow in the measuring circuit, and the amount increases as the transducer resistance increases. The effect of this shunting is to decrease the amplitude of the change in transducer resistance; hence, the measured value of ΔV (see Fig. 21) will be in error. Keough (1968) discussed the problem of data reduction and the method of correcting for the effects of non-constant current and shunting by the measuring circuit.

For transducer resistances of the order of 50 Ω, the corrections to the data are large. Another approach to the use of such transducers has been to use a bridge circuit with a constant voltage power supply to measure the transducer output (see Keough, 1963). The bridge circuit and power supply developed at SRI is shown in Fig. 22. Similar circuits have been used by other workers (Kusubov and van Thiel, 1969). A constant voltage is applied between points A and B in Fig. 22, and oscilloscopes are used to measure the voltage between points C and D. Resistors R_1 and R_2 form one bridge leg, and R_3 and the transducer form the other leg. The bridge is pulsed and R_1 adjusted until no voltage is detected between C and D, i.e. the bridge is balanced. When the transducer is stressed, its resistance changes and the bridge becomes unbalanced. This imbalance is measured by oscilloscopes as a voltage between C and D. To calibrate the bridge, an accurately known amount of resistance (2 to 10 Ω) is temporarily placed in series with the transducer after the bridge is balanced. The power supply is discharged

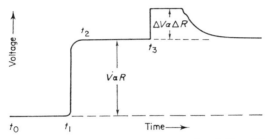

Fig. 21. Idealized ocsilloscope waveform from circuit of Fig. 20. (Power supply triggered at t_0; current turn-on initiated at t_1 and completed at t_2; stress signal begins at t_3.)

and the voltage between C and D measured. If the resistance of the transducer is accurately known, this procedure is sufficient to calibrate the bridge for the specific transducer being used.

Stress-resistance calibrations of manganin have been obtained by embedding it in a material with a known equation of state and subjecting the material to a known stress. The output of the transducer is measured and associated with that particular stress. The assumption is made that the active element comes to stress equilibrium with the surrounding material. In fact, the rise time or response time of the transducer is proportional to the wire thickness and depends also on the impedance mismatch between the active element and the surrounding material (It should be pointed out that the transducer measures only the normal component of the compressive stress.) In practice, it is $\Delta V/V$ that is measured (see Fig. 21), and the stress is proportional to $\Delta R/R$ where ΔR is the resistance change with stress and R is the initial transducer resistance. Care must be taken in relating $\Delta V/V$ to $\Delta R/R$

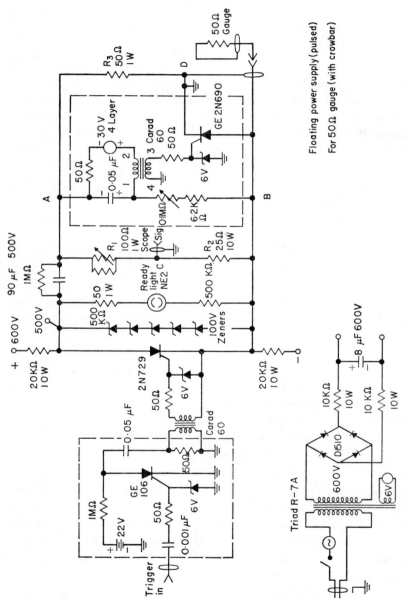

Fig. 22. Constant-voltage pulsed power supply with bridge circuit to measure transducer output.

for the particular circuit used. The circuit parameters that influence the observed $\Delta V/V$ are (1) the shunt resistance of the voltage measuring circuit and (2) the changing current through the transducer as the change in transducer resistance becomes an appreciable percentage of the ballast resistor (for constant current circuits). A complete discussion of these factors and the way in which corrections can be applied has been given by Keough (1968).

The dynamic piezoresistance function for manganin has been measured by several workers with some disagreement. Those, measurements known to the authors are given in Table I. Keough (1968) made measurements above 170 kbar and found a change in the piezoresistance coefficient. For stresses above 170 kbar

$$\sigma(\text{kbar}) = (\Delta R/R - 0.16) \ [(0.18 \pm 0.01) \times 10^{-2}]^{-1}, \qquad (93)$$

whereas between 20 and 170 kbar

$$\sigma(\text{kbar}) = (\Delta R/R) \ [(0.29 \pm 0.01) \times 10^{-2}]^{-1}. \qquad (94)$$

Lyle and Schriever (1969) were able to fit their data with a polynominal of the form

$$\sigma(\text{kbar}) = \alpha(\Delta R/R) + \beta(\Delta R/R)^2 + \gamma(\Delta R/R)^3 \qquad (95)$$

with $\alpha = +3.70 \times 10^2$, $\beta = -1.32 \times 10^1$, $\gamma = +4.42 \times 10^1$. The exact reason for the discrepancy in the relationship between stress and resistance change determined by various workers is not known, but several factors can be suggested. The composition of the manganin used was slightly different, and the design and emplacement of the transducers was different. The last two factors may well be the major ones. Keough and Wong (1970) investigated the piezoresistance coefficient of manganin in wire and thin-foil geometries in high-density polycrystalline alumina (Lucalox). They report that for Lucalox shocked below the Hugoniot elastic limit and containing a transducer in the wire geometry, this coefficient was much greater than in C-7 epoxy or in Hi-D glass. For stresses above the HEL of Lucalox (~ 100 kbar), the coefficient of manganin agrees well with that of manganin wire in other insulators. When a transducer in the thin-foil geometry was used in Lucalox, the coefficient was the same above and below the HEL. Keough and Wong postulated that the higher coefficient was the result of lattice defects produced in the manganin by plastic deformation of the wire compressed at stresses less than the dynamic yield stress of

TABLE 1

Dynamic piezoresistance coefficient of manganin

Source	Manganin composition	Average piezoresistance coefficient (Ω/Ω) kbar^{-1}	Stress range kbar	Accuracy %
Bernstein and Keough (1964)[a]	87% Cu, 13% Mn	0.00293	56 to 145	±3
	85.69% Cu, 10.4% Mn 3.66% Ni, 0.15% Fe 0.03% Si, 0.06% Co 0.001% Ag, <0.01% Al	0.0025	56 to 145	±3
Keough (1968)[a]	86% Cu, 12% Mn, 2% Ni	0.00292	56 to 145	±3
Fuller and Price (1969)[b]	84% Cu, 12% Mn, 4% Ni	0.0029	20 to 170	±3
Kusubov and van Thiel (1969)[c]	86% Cu, 12% Mn, 2% Ni	0.002363	to 300	—
	—	0.0026	to 130	±8
Lyle and Schriever (1969)[d]	83% Cu, 13% Mn, 4% Ni	0.00264	77 to 132	±3

[a] Calibration based on C–7 epoxy Hugoniot.
[b] Calibration based on Araldite epoxy Hugoniot.
[c] Calibration based on 2024-T4 aluminium Hugoniot.
[d] Calibration based on copper and aluminium Hugoniots.

the Lucalox. These results indicate that careful consideration should be given to the mode of compression that the manganin will experience when used as a stress transducer. Keough and Wong (1970) suggest that foils with a large aspect ratio (e.g. 100 to 1) be used in stress transducers. Barsis *et al.* (1970) have also studied the piezoresistivity of manganin as a function of the mode of strain. Their results, like those of Keough and Wong (1970), indicate that the manner in which the transducer element is strained is important.

Stress transducers can be used in several ways. The transducer may be embedded in an insulator, generally a castable epoxy resin. This block of resin containing the transducer is then placed on the material to be studied so that the plane of the shock front and that of the transducer element are parallel. The stress profile is measured in the epoxy resin, and the profile in the sample material is then found by the impedance match method. This transducer geometry has been used by several workers (see Bernstein and Keough, 1964; Fuller and Price, 1964; Keough *et al.*, 1964; and Keough, 1968). The transducer may be embedded directly in the material to be studied. If the material of interest is a conductor, the transducer must be electrically insulated. Thin layers of epoxy, glass, mica, and plastic have been used for this insulation. The transducer and insulation must be as thin as possible to minimize the transducer rise time.

Embedding the transducer directly in the material of interest has an important advantage in that the stress–time profile obtained is the profile in that material and is not distorted by interactions with a back-surface transducer, i.e. no impedance matching is necessary. For examples of this implacement geometry, see Fuller and Price (1969), Kusubov and van Thiel (1969), Lyle and Schriever (1969), and Keough and Wong (1970).

The manganin stress transducer has been used to measure compression profiles and the release profiles. As discussed previously, the transducer measures the stress component normal to the front of the stress wave. There is currently no method of measuring the lateral stresses associated with the stress wave. One attempt has been made by Bernstein *et al.* (1968) to use a transducer to measure the deviatoric stresses in a material during the passage of a stress wave. Their effort was not entirely successful, and apparently there are fundamental problems with transducer emplacement for this type of measurement.

At low kilobar stresses, the change in transducer resistance is small

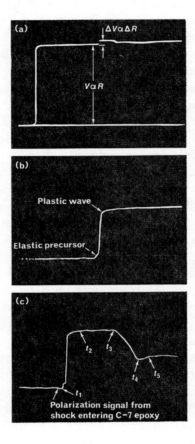

Fig. 23. Manganin transducer voltage–time records from an aluminium target impacted by an aluminium projectile. The transducer was embedded in C–7 epoxy and mounted on the target surface opposite the impact. The manganin wire was about 1 mm from the aluminium surface. Peak stress in the aluminium is 17 kbar. (a) Oscilloscope trace displaying current turn-on pulse and the change in voltage due to stress. $\Delta V/V$ ratio for peak stress was obtained from this oscilloscope. (b) Oscilloscope trace displaying the stress wave only. Note the elastic precursor. Time between dots is 0.5 μs. (c) Oscilloscope trace displaying detail of the stress wave. Time between dots is 0.5 μs. t_1 is arrival of elastic precursor at the manganin wire. t_2 is the peak of the plastic wave of 17 kbar. t_3 is the arrival of the unloading wave from the free surface of the projectile. t_4 is the arrival of the minimum stress portion of the unloading wave at the target–transducer interface. t_5 is the arrival of a compression wave caused by separation of target and projectile.

and some difficulty in recording the stress–time profile can occur. For example, with a 1-Ω transducer and a current of 5 A, a stress of 10 kbar would produce a voltage step ΔV of 0.145 volts. To measure 0.145 volts on top of the current step V of 5 volts, two methods have been used. Generally, an offset preamplifier (such as Tektronix Type W or 1A5) is used, and a negative voltage equal to the turn-on voltage is applied to cancel the current step and display only the stress signal ΔV. In Fig. 23 are shown the transducer voltage-time records from an experiment in which an aluminium target was impacted by an aluminium projectile (Keough, 1968). The transducer was embedded in a C–7 epoxy block backing the target. The manganin wire is about 1 mm from the aluminium C–7 interface. The use of the offset preamplifier in the manner just described has the disadvantage of baseline drift. This drift originates in the preamplifier and also from the normal inconsistencies associated with repetitive discharge of the power supply.

Two other techniques have been used at SRI to record low kilobar stresses. In one, a high-quality delay line is used, and the offset feature is not needed. In the other, the offset preamplifier is used with an auxiliary circuit in a manner that minimizes any drift.

The delay line circuit designed by W. G. Ginn (see Keough, 1968) is shown in Fig. 24(a) and an idealized waveform in Fig. 24(b). Current flow is initiated at time t_0; at t_2, which is determined by the length of the delay cable, the current step is applied to the minus B input of a suitable differential preamplifier (Tektronix W or 1A5). Thus B is subtracted from A, and the system is in a quiescent state. The voltage sensitivity of the preamplifier can be set so that small voltages produce a large deflection. Thus a small change in transducer resistance, i.e. a small stress pulse, will appear as a signal on A and as a delayed negative signal on B. The timing of the stress wave is arranged so that the stress signal occurs after the quiescent state described above has been established. The recording duration is limited by the length of the delay line at B ($5\mu s$ in the circuit of Fig. 24(a)). The oscilloscope is normally triggered after t_3 and before t_4 and therefore views only the stress signal.

Figure 25 shows a circuit designed by Ginn, Hall, and Keough (private communication) using a differential offset technique. In using this circuit, one leg of the power supply is grounded as shown in Fig. 25. A Tektronix type 1A5 or type W preamplifier is used and must be operated in the d.c. mode, and the display is set at A–B. The transducer turn-on voltage is balanced out using the comparison voltage monitor

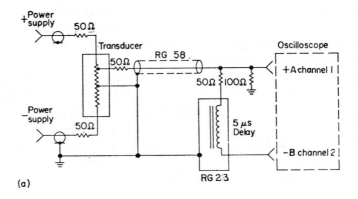

(a)

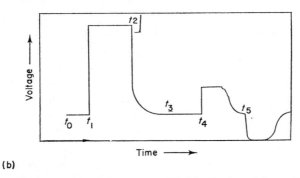

(b)

Fig. 24. (a) Delay-cable recording system. (b) Idealized waveform of the oscilloscope voltage of Fig. 21. t_0, Trigger applied to power supply; t_1, current turn-on arrival at channel 1 (see above); t_2, delayed current step arrival at channel 2; t_3, quiescent state; t_4, stress signal arrival at channel 1; t_5, stress signal arrival at channel 2.

output from the preamplifier. The comparison voltage is mixed with the transducer turn-on voltage and applied to the preamplifier attenuators at the same point as the transducer signal. Thus the turn-on voltage is balanced out regardless of the vertical deflection setting of the preamplifier.

A different approach to obtaining wave profiles at low stresses is to substitute a more sensitive piezoresistive material for manganin. Those materials investigated thus far are listed in Table 2. Ytterbium and carbon are the most promising materials at present. Ytterbium has been used recently at SRI and elsewhere in stress transducers for gas-gun experiments and for field experiments near an

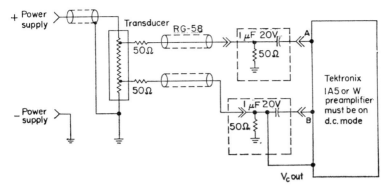

Fig. 25. Differential offset recording circuit.

TABLE 2

Dynamic piezoresistance coefficients

Element	Average piezoresistance coefficient $(1/R)$ $\Delta R/\sigma$ (Ω/Ω) kbar^{-1}	Stress range kbar	Reference
Calcium	0.023	up to 27	Williams and Keough (1966, 1967)
Lithium	0.023	up to 54	Williams and Keough (1966, 1967)
Cadmium	0.0058	at 28	Williams and Keough (1966)
Indium	0.0028	at 15	Williams and Keough (1966)
Lead	0.0033	15 to 19	Williams and Keough (1966)
Bismuth	0.033	at 10[a]	Keough (1970)
	0.045	at 20	Keough (1970)
Ytterbium	0.055	at 4[b]	Keough (1970)
	0.127	at 21	Keough (1970)
Carbon	−0.030	up to 15[b]	Ginsberg (1973, private communication)

[a] Bismuth has nonlinear response; stress values given are equilibrium stress states.

[b] Ytterbium and carbon have nonlinear response; stress values given are equilibrium stress states.

explosive source. Ginsberg (1971, 1973) has performed a number of experiments calibrating ytterbium in loading and unloading up to 33 kbar. It undergoes a transition from the semiconductor to the metallic state near this stress value. He has also investigated a number of the phenomena that affect the response of piezoresistive gauges, such as the metallurgical history of the foil used as the active element, and the effect of multidimensional strain fields. Most of the measurement techniques described above for manganin also apply to ytterbium, and ytterbium has a number of advantages for use in high-noise environments, and in divergent flow fields such as those encountered in *in situ* geological measurements. It has proved itself to be especially useful in inhomogeneous materials such as composites. Its small response to local deformation, compared to its overall sensitivity, makes it quite attractive in this application. Carbon gauges have been used to make stress measurements in radiation environments, and are also gaining acceptance for use in laboratory experiments. Sensitivity approximates that of ytterbium at lower stresses, but levels off rapidly at stresses of 50 kbar and over. Although carbon has not been studied as extensively as ytterbium or manganin as a transducer material, it is a useful addition to this class of instrumentation.

The experimenter has a good deal of latitude in selecting the type (two- or four-lead) and amount of resistance of the transducer to be used. However, there are some important factors to be considered in making this choice. First, low-resistance two-lead transducers are more susceptible to contact resistance than low-resistance four-lead ones. Thus a four-lead low-resistance system is a better choice for low resistance transducers (1 to 2 Ω). Second, large resistance transducers are more sensitive, which is an important consideration when stresses less than 20 kbar are to be encountered. Third, the type of transducer power supply must be considered. The constant current supply can be used for any resistance transducer, but if the resistance is larger than 1 to 2 Ω, corrections must be made to obtain the true $\Delta V/V$ as discussed above. For large resistance transducers (5 to 50 Ω), a bridge circuit with a constant voltage power supply is a better choice. When using the bridge circuit, a two-lead transducer is used, but since the resistance is normally large, any contact resistance will be a negligible part of the total resistance. A final consideration is the physical size of the transducer; the smaller the transducer, the less sensitive to shock tilt it will be.

Noise considerations in instrument cables and experimental environment have been discussed by Perls (1952), Linde and Schmidt (1966b), and Williams (1968).

5.1.3 Electromagnetic transducer

Young et al. (1970) recently developed an electromagnetic stress trans-ducer. In Fig. 26 a schematic view of this transducer is shown. The use of this transducer is based on the equation of motion for waves in Lagrangian coordinates and on Faraday's law for electromagnetic

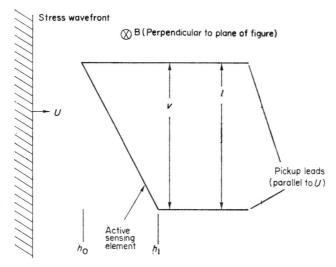

Fig. 26. Schematic view of electromagnetic stress gauge. (From Young et al., 1970.)

induction. The equations expressing conservation of mass and momen-tum in Lagrangian coordinates can be combined to give (see Cowper-thwaite and Williams, 1971)

$$\rho_0 \left(\frac{\partial u}{\partial t} \right)_h + \left(\frac{\partial \sigma}{\partial h} \right)_t = 0, \tag{96}$$

where the symbols used here are those previously defined. Integrating equation (96) with the limits of integration as shown in Fig. 26, we obtain

$$\sigma(h_1, t) - \sigma(h_0, t) = -\rho_0 \frac{\partial}{\partial t} \int_{h_0}^{h_1} u \, dh. \tag{97}$$

We note that the integral represents the total momentum between h_1 and h_0, and this is the quantity that the transducer measures.

In Fig. 26 the wire or foil element of the transducer is inclined at an angle θ to the wavefront, and the magnetic field is perpendicular to the plane containing the transducer. For a constant magnetic field B, the voltage output produced by particle motion (transducer moves with the particles) perpendicular to the magnetic field is given by Faraday's law

$$V = -B \frac{dA}{dt} \tag{98}$$

where A is the area enclosed by the transducer loop. As the stress wave encounters the sensing element, the material moves with a particle velocity u, and the change in loop area with time (see **Fig. 27**) is given by

$$dA = -dx\, dy \tag{99}$$

with $dx = u\, dt$, and

$$y = \frac{l(h - h_0)}{h_1 - h_0}.$$

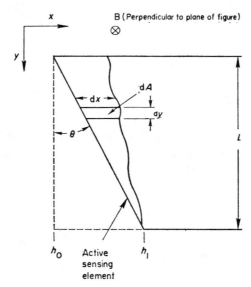

Fig. 27. Change in loop area as stress wave moves across the transducer (From Young *et al.*, 1970.)

Thus

$$dy = \frac{l \, dh}{h_1 - h_0}$$

and

$$dA = \frac{-l}{h_1 - h_0} u \, dh \, dt,$$

and finally

$$\frac{dA}{dt} = -\frac{-l}{h_1 - h_0} \int_{h_0}^{h_1} u \, dh. \tag{100}$$

Therefore the voltage output of the transducer is

$$V = \frac{Bl}{h_1 - h_0} \int_{h_0}^{h_1} u \, dh. \tag{101}$$

From equation (97) we can write

$$\sigma(h_1, t) - \sigma(h_0, t) = -\frac{\rho_0(h_1 - h_0)}{Bl} \frac{dV}{dt}. \tag{102}$$

The transducer has a continuous output, and if h_1 is a free surface the derivative of this output is proportional to the stress at h_0. If a series of pickup leads are used as shown in Fig. 28, stress–time profiles

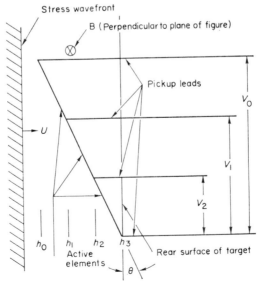

Fig. 28. Electromagnetic stress gauge with stress readout at three depths and with reference pickup h_3 on free surface. (From Young et al., 1970.)

can be obtained at several locations. An important feature of this transducer is that it requires no prior calibration. Transducer emplacement, the restriction of its use to nonconducting specimens, and the necessity of taking the derivative of an experimental curve are some disadvantages.

5.1.4   *Anomalous thermoelectric effect*

Crosnier *et al.* (1965) attempted to use the so-called "anomalous thermoelectric effect" in shock-loaded metals as a stress transducer. This effect (discussed in some detail in section 6.1.3) occurs when a stress wave crosses the interface between two different metals. A voltage that depends on the stress and the particular metals in contact appears between the uncompressed extremities of these metals, see Fig. 29.

Crosnier and co-workers showed that the stress at a copper–constantan junction is linearly related to the observed voltage between 180 and about 1600 kbar. They report an error of less than 2 per cent between

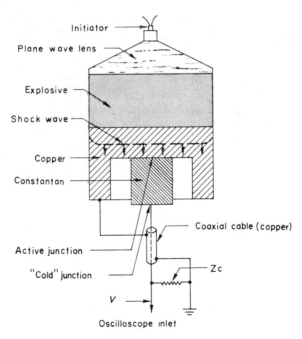

Fig. 29. Basic thermocouple configuration for observing anomalous thermoelectric effect. (From Crosnier *et al.*, 1965.)

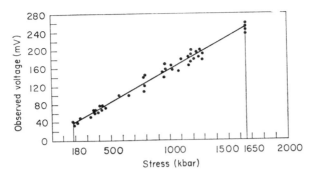

Fig. 30. Experimental curve of voltage versus shock stress for a constantan–copper junction. (From Crosnier *et al.*, 1965.)

200 and 1200 kbar. Their results are shown in Fig. 30 and can be represented by the expression

$$V(\text{millivolts}) = 0.45 \; \sigma \; (\text{kbar}) + 11.$$

More recent work was summarized by Baum *et al.* (1973).

5.1.5 *Polar solid transducer*

When a polar solid is shock loaded, a voltage appears between the surfaces of the solid. The experimental arrangement is shown in Fig. 31. This "shock polarization" of nonconducting materials is discussed in detail in section 6.1.2. Eichelberger and Hauver (1962) first investigated this phenomenon and have used it to study the shape of the stress wavefront in TNT. Their transducer's configuration is shown in Fig. 32.

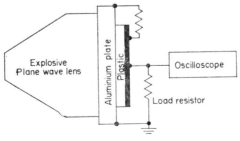

Fig. 31. Experimental setup for investigating shock polarization of polar solids. (From Eichelberger and Hauver, 1962.)

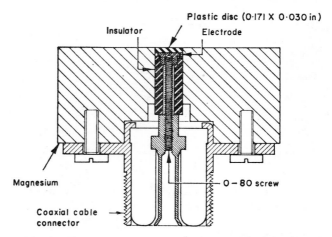

Fig. 32. Transducer design using a polar solid. (After Eichelberger and Hauver, 1962.)

This is the only application of a transducer based on shock polarization of which the authors are aware.

5.2 PARTICLE VELOCITY MEASUREMENT

5.2.1 *Free-surface measurements*

Measurements of particle velocity can be divided into two general types, free surface and in-material measurements. Most in-material techniques, of course, can also be used for free-surface measurements. Many techniques have been developed for measurement of free-surface velocity, mainly by optical or electrical means. These methods are reviewed in detail by Doran (1963) and by Duvall and Fowles (1963).

Recent developments include the use of an inclined prism instead of an inclined mirror to obtain greater sensitivity in free-surface velocity measurements (Eden and Wright, 1965). Light enters one side of the prism (Fig. 33), reflects from the bottom by total internal reflection, and exits from the opposite side. As the shock front impacts the prism bottom (Fig. 34), the total internal reflection quality is destroyed. This technique gives sharper light cutout at lower stress levels than does the inclined mirror technique. In Fig. 34 the arrangement of prisms to measure the free-surface velocity of a specimen is shown.

A disadvantage of free-surface velocity measurements is that they are not related merely to in-material velocities. The free-surface motion depends on both the dynamic state in the material and the particular

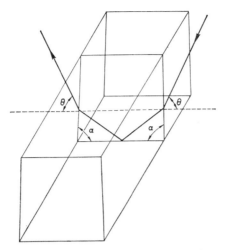

Fig. 33. Light path in transparent prism.

properties of the material. In some cases the approximation that the free-surface velocity is twice the particle velocity is sufficient. In other cases an iteration procedure is used to obtain better agreement (Barker *et al.*, 1964). For porous materials, the free-surface velocity can be much less than twice the particle velocity (Petersen *et al.*, 1970). If other parameters are measured in addition to free-surface velocity, the results may be used to study the dynamic stress–strain relations of the material (Munson and Barker, 1966).

5.2.2 *In-material measurements*

In the past few years much progress has been made in the development and use of in-material gauges. For particle velocity measurements, these include the electromagnetic gauge and the laser interferometer.

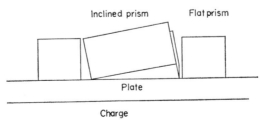

Fig. 34. Free-surface velocity measurement in plate using prisms.

a. *Electromagnetic gauge*   The electromagnetic gauge is based on the principle of a moving conductor in a magnetic field. If a straight wire conductor of length l is moved at a velocity u through a uniform magnetic field H, a voltage V given by

$$V = (\vec{l} \cdot \vec{u} \times \vec{H}) \tag{103}$$

will be induced.

This technique was first reported for measuring particle velocities in detonations (Zaitsev *et al.*, 1960; Dremin and Shvedov, 1964; Jacobs and Edwards, 1970). The technique has also been used to measure particle velocity during compression of various nonconducting materials (Dremin and Adadurov, 1964; Dremin *et al.*, 1965; Frasier and Karpov, 1965; and Ainsworth and Sullivan, 1967). Al'tshuler *et al.* (1967) and Petersen *et al.* (1970) measured particle velocities during release from the shocked state.

The electromagnetic gauge is a metal wire or foil embedded in the medium to be studied. The specimen is placed in a uniform magnetic field so that the field lines, the direction of shock propagation, and the active gauge element l are mutually perpendicular (Fig. 35). The leads are brought out parallel either to the field or to the direction of shock propagation so that the voltage induced by the motion of the material is influenced only by the active element and is directly proportional to the particle velocity. If precautions are taken to protect the leads, particle velocity can be monitored continuously during compression and the subsequent release (Petersen *et al.*, 1970) (see Fig. 36).

Also shown in Fig. 35 are the projectile and target for a gas-gun experiment. A Lucite extension is mounted on the aluminium projectile to prevent a disturbance to the magnetic field caused by a conductor moving into the field. In the experiment shown, the impacting head on the Lucite extension is the same material as the target. The impact is then symmetrical, with the particle velocity behind the shock in the target equal to half the projectile velocity. In most experiments using this technique, the measurement of projectile velocity was considered more accurate than the combination of measurements of effective element length and magnetic-field strength. Therefore, the steady voltage induced by passage of the shock is taken as representing a particle velocity equal to half the projectile velocity. The measurement during compression and release can be related to this particle velocity as ratios of induced voltage. Magnetic field and foil length

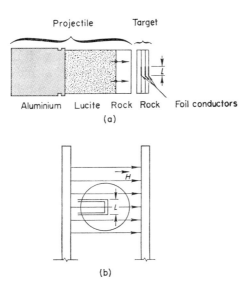

Fig. 35. Particle-velocity gauge. (a) Side view of projectile and layered target with three foils mounted at interfaces. (b) Front view showing position of a foil at a target interface. Target is mounted between poles of a magnet. In general, the effective length of the gauge element, to be used in calculating the voltage produced by particle motion, will be less than the length l shown above. (After Petersen *et al.*, 1970.)

should always be measured as a check in symmetrical impact experiments. If symmetrical impact is not practical as in explosive experiments, particle velocity is determined directly from equation (103).

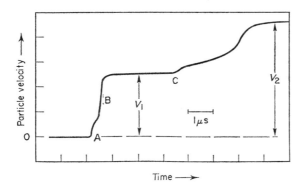

Fig. 36. Particle velocity gauge record in Solenhofen limestone. Precursor arrives at A, main wave at B, relief wave from free surface at C. Note the spreading of the relief wave. Voltages induced in gauge by particle velocity in shocked state and by free-surface velocity are given by V_1 and V_2, respectively.

The particle velocity–time profile from an impact experiment with Solenhofen limestone is shown in Fig. 36.

b. *Interferometer techniques*   Laser interferometer techniques can be used to monitor displacement or velocity, either of a reflecting free surface or of a reflecting interface between a shocked sample material and a transparent "window" material. Sensitivity can be very high. Shock rise times down to about 1 ns have been resolved, yet total recording times of many microseconds are possible (Barker and Hollenbach, 1970).

A limitation of the method is that the surface to be observed must remain reflecting under shock loading. Fortunately, this condition can be met for many materials for stress regions of interest. Another limitation is that true in-material measurements can be taken only for transparent materials, where a reflecting surface between two pieces of the specimen material can be observed. Otherwise the observations are made at a free-surface or at an interface between the specimen material and a transparent window. If the window material and the specimen have similar shock impedances, the observation will only be slightly affected by stress-wave reflections at the interface.

Two general types of laser interferometers are currently being used in shock-wave experiments. At low velocities a Michelson interferometer can be used (Barker and Hollenbach, 1965). A diagram showing its operation is given in Fig. 37. As the observed surface moves in response to the arrival of a stress wave, the interferometer produces light fringes in proportion to the displacement. The fringe frequency is proportional to the reflecting surface velocity (see Fig. 38). At high surface velocities, the fringe frequency can exceed the response capability of the photomultiplier tube. This and other frequency-related problems limit the Michelson interferometer technique to the measurement of velocities less than about 0.2 mm μs^{-1} (Johnson and Barker, 1969).

For higher velocities, the velocity interferometer may be used. Figure 39 shows the experimental setup as given by Barker (1968). A single-frequency He–Ne laser beam is focussed on the mirror surface at the target–window interface. The reflected beam is recollimated by the same lens and is then sent through the velocity interferometer, which consists of two beam splitters and a 90° reflecting prism. Motion of the reflecting interface produces a Doppler shift in the reflected light frequency, and the velocity interferometer beats the current light

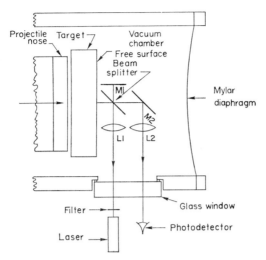

Fig. 37. Schematic of the interferometer configuration for measuring the free surface motion of the target specimen. M1 and M2 are mirrors. The lens L1 focuses the laser beam to a point on the free surface, and the lens L2 partially recollimates the light. (After Barker and Hollenbach, 1965.)

frequency against the frequency existing a few nanoseconds earlier by sending part of the light around the delay leg (Johnson and Barker, 1969). In this way, light fringes are produced at the photomultiplier tube with the number of fringes proportional to the interface velocity and the fringe frequency proportional to the acceleration. Generally, the number of fringes can be estimated to better than ± 0.1 fringe, giving an interface velocity accuracy of better than 1 per cent if a total

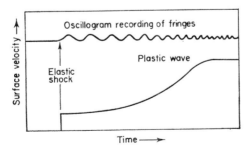

Fig. 38. Oscillogram of fringes from the Michelson interferometer and corresponding specimen surface velocity history. A relatively low velocity experiment is shown here. At velocities above about 0.2 mm μs^{-1} the fringer frequency is too high to be tracked by the photomultiplier tube. (After Johnson and Barker, 1969.)

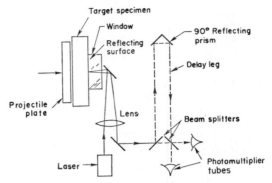

Fig. 39. Experimental setup for velocity interferometer system. (After Johnson and Barker, 1969.)

of ten fringes is recorded. Figure 40 shows an oscillogram trace of the photomultiplier tube output together with the corresponding velocity history.

A detailed discussion of the analysis of such records was given recently by Clifton (1970b). Clifton analyses the change of phase at the photomultiplier in Fig. 39 when the target free surface is accelerated. This change of phase will cause the photomultiplier to record a number of fringes $\Delta N(t)$, and Clifton shows that the velocity of the target surface is given by

$$u(t - \tau/2) = (\lambda/12\tau) \, [5\Delta N(t) + 2\Delta N(t - \tau/2) - \Delta N(t - \tau)],$$

$$t > 2\tau, \quad (104)$$

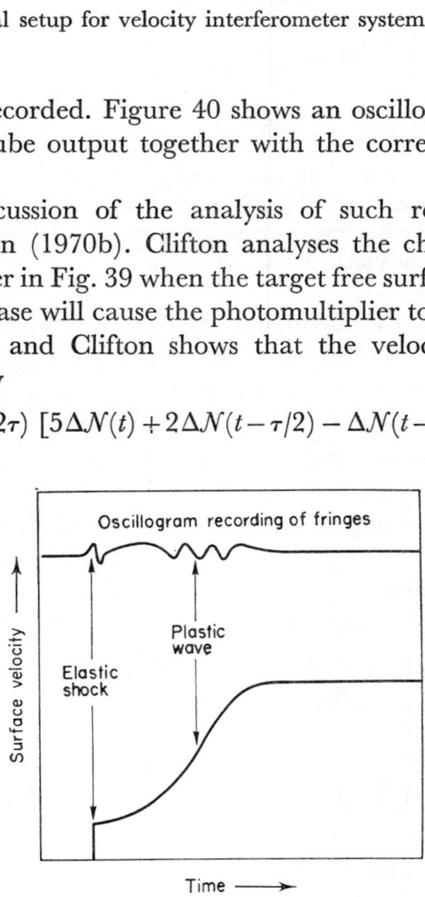

Fig. 40. Oscillogram of fringes from the velocity interferometer and corresponding specimen surface velocity history. (After Johnson and Barker, 1969.)

where τ is the time required for light to travel around the delay leg in the interferometer. The formula is exact for motions for which the jerk d^2u/dt^2 is constant. Clifton also shows that in practice very little error is introduced by using

$$u(t-\tau/2) = (\lambda/2\tau)\,\Delta\mathcal{N}(t) \tag{105}$$

to obtain the target surface velocity. Formula (105) is essentially the relation that was used by previous investigators and holds strictly only if the acceleration is constant during the time τ.

Another higher-order evaluation of such interferometer records has recently been given by Barker (1971b).

If a window material is used, as is shown in Fig. 39, the analysis is complicated by the effects of the stress wave on the index of refraction of the window. The index of refraction and its variation with density associated with the transmitted stress must be accounted for in a new calculation of the proportionality between velocity and number of fringes. Also the dynamic material properties of the window must be known to analyse the experimental results properly. Polymethyl methacrylate (PMMA), fused silica, and sapphire were studied by Barker and Hollenbach (1970) for use as window materials.

Johnson and Burgess (1968) discuss a velocity interferometer technique that differs in two important respects from that of Barker (1968), which has been presented so far. Johnson and Burgess have a lensless optical system that permits the monitoring of a high-velocity, reflecting surface without tilt-induced measurement inaccuracy. Also they use Gilson's (1967) mode-matching scheme that permits the use of high-power, multilongitudinal mode lasers for applications where the required frequency resolution is well within the longitudinal mode separation. Thus the use of a single frequency laser with the acceptance of low optical power is no longer required. Supplementary features of the system presented by Johnson and Burgess resulting from the use of the mode matching scheme and high-power lasers are: (1) diode rather than multiplier tube photodetectors are permitted, and (2) reflectivity requirements of the free surface may be relaxed.

Barker and Hollenbach (1972) have developed a laser interferometer system to measure the velocity history of either spectrally or diffusely reflecting surfaces. The system produces two interferometer fringe signals in quadrature to improve resolution and to distinguish acceleration and deceleration. A very large depth of field can be achieved with this system and in this respect the velocity history of a projectile during its acceleration down a long gun barrel was measured. Barker (1972) has reviewed the use of laser interferometry in shock-wave research.

5.2.3   Flash X-ray measurements

Flash X-ray techniques have been used for many years for observing high-speed phenomena associated with shock and detonation waves. Venable and Boyd (1965) provide many of the early references. The main attraction of flash radiography is that of direct observation of an explosive event. In certain cases, especially in studies of explosives, X-ray techniques can be used to measure particle velocities.

By embedding a series of foils at precisely known initial positions in a low-density material to be studied and allowing the shock wave (or detonation wave in an explosive) to sweep by these foils, flash radiography can be used to measure foil displacement as a function of time (Venable and Boyd, 1965). This displacement can be related to particle velocity behind the front by the theory being examined. The quality of fit of these data thus provides the means of testing the applicability of the theoretical model.

In general the X-ray techniques lack the precision of other methods of measuring particle velocity. Detail is lost because of the finite exposure time required and because continuous observation as a function of time cannot be made. On the other hand, a very powerful use of flash radiography for observation of multiple spall has been presented by Breed et al. (1967). Both the layer thicknesses and their space–time positions are determined. By virtue of this unique capability, the model used for data interpretation can be tested. Another application, the observation of the formation of the first and second plastic waves in a shock-induced phase change in antimony, is demonstrated by Breed and Venable (1968). The time required to complete the phase transformation behind the first plastic wave is shown to be 2 to 3 μs. In earlier work in iron, Balchan (1962) successfully observed the double shock structure and the rarefaction shock.

5.2.4   Immersed foil technique for release measurements

The immersed foil technique (Ahrens and Ruderman, 1966) was developed to determine points on the release curves of solids. The

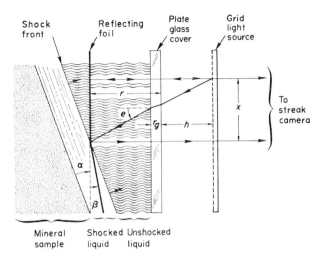

Fig. 41. Diagrammatic cross section, immersed foil release adiabat experiment. Upon interaction with the shock wave, the reflecting foil (initially at an angle α to the sample surface) is rotated through an angle β. This rotation with resulting light-ray refraction and reflection effects causes light-source image to be displaced a distance x, which is recorded by camera.

reflecting foil is mounted in a transparent liquid, close to and at an angle to the liquid–solid interface (Fig. 41). A streak camera views the displacement of the reflection of a grid as the shock, which propagates through the solid, encounters the foil (Fig. 42). The magnitude of the displacement is related to the angle by which the foil is rotated by the shock, which is, in turn, related to the particle velocity. The rate at which the shock intersects the inclined foil is related to shock velocity.

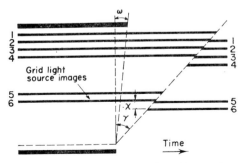

Fig. 42. Diagram of streak camera record, showing film angle and image displacements resulting from shock front propagating within immersed foil assembly.

These values give the shocked state for the liquid, which is a point on the stress–particle velocity release curve of the solid. A correction must be made for the change in refractive index of the shocked liquid. The advantages of the method are that states corresponding to two (or more) shock fronts can be determined and that a precise prior knowledge of the liquid Hugoniot is not required because both shock velocity and particle velocity are determined during the experiment.

The shock velocity in the liquid is determined by the velocity with which the shock front intersects the inclined foil. The shock velocity is given by the formula

$$U_1 = \frac{W}{M} \frac{\sin \alpha}{\tan \gamma - \tan \omega}, \tag{106}$$

where W is the streak camera writing rate on the film, M is the magnification from the assembly to the film in the plane of the foil, γ is the cutoff angle on the film, and ω is the shock tilt.

The particle velocity in the liquid is obtained by measuring the grid image displacement on the film. This displacement occurs because the foil turns through an angle β when it achieves the particle velocity of the surrounding shocked liquid. The success of the particle velocity determination depends upon the thin foil reaching the particle velocity of the surrounding liquid in a time that is short compared with the resolution time of the streak camera (10^{-8} s) without losing its reflectivity. Under these conditions the foil behaves like a particle or slab of liquid whose motion can be directly monitored. The displacement x of the grid image on the film is related to the particle velocity u in the fluid in terms of geometrical parameters of the experimental assembly, α, h, r, and r_g (Fig. 41), the indices of refraction of the unshocked and shocked liquid n_0 and n_1, the index of refraction of the glass, and the grid image magnification. Independent experiments must be performed to obtained the refractive index of the shocked liquid.

The immersed foil technique has been largely replaced by the use of multiple in-material gauges for measuring loading and release paths (see below section 5.6).

5.3   MEASUREMENT OF WAVE VELOCITY

Shock velocity is the easiest quantity to measure in dynamic experiments. With detectors at two different levels, one simply records the time difference between the arrival of the event at detectors separated by a known distance. These can be time-of-arrival gauges, such as charged coaxial cables (Ingram, 1965), flat reflecting mirrors (Doran, 1963), argon flash gauges (Walsh and Christian, 1955), plastic tapes that emit current pulses when impacted by strong shocks (Champion and Benedick, 1968), anodized aluminium switches (Lysne, 1968), and a number of other techniques for detecting the arrival of a shock at a surface.

If the stress wave does not arrive as a single sharp front or if more detailed information about the pulse shape is desired, the above techniques may not be adequate. Usually the wave velocity measurement is made as part of the measurement of stress or particle velocity. A time-of-arrival detector is placed at the explosive or impact interface and is correlated in time with the other gauge or gauges so that the time for the wave to propagate through the specimen is recorded. In this way the development of any structure in the stress wave is also recorded. With in-material gauges at different planes, wave velocity is obtained as a function of stress or particle velocity, or both.

5.4 MEASUREMENT OF SPECIFIC VOUME

The most difficult measurement to make is that of specific volume. Dapoigny *et al.* (1957) and Schall (1958) used flash X-ray techniques to determine density directly. Unfortunately, in one-dimensional strain experiments rarefactions are generated at the boundaries, and these interfere with X-ray measurements. Flash X-ray techniques have been very successful for more specialized problems, such as the direct observation of multiple spalling (Breed *et al.*, 1967).

A recent advance, reported by Q. Johnson *et al.* (1970), is the observation of X-ray diffraction during shock-wave compression. Timing presents serious difficulties, but the technique permits the direct observation of the crystallography of dynamic phase changes. The initial experiments of Johnson *et al.* with LiF at about 130 kbar and those of Egorov *et al.* (1972) on aluminium at 140 kbar have shown that crystalline order can exist behind the shock front and that the ordering takes place on a time scale short in comparison with 20 ns. The shock compression is shown to be essentially hydrostatic. Johnson and Mitchell (1972) have obtained the first X-ray pattern of a solid at the instant it is undergoing a phase transformation. They report on measurements

5

of pyrolytic boron nitride transforming to a high-pressure phase which they suggest has a wurtzite phase. These new developments should make it possible to determine the volume of materials and to study polymorphism in materials during shock-wave loading.

Two types of X-ray detector systems have been developed for use with flash X-ray studies of shock-compressed materials. A scintillation detector system has been described by Johnson *et al.* (1971) and more recently a film system has been described by Mitchell *et al.* (1973).

5.5   MEASUREMENT OF SHOCK TEMPERATURE

By measuring the parameters of a shock wave, the velocity U and the particle velocity behind the shock front u, the stress, volume, and increase in internal energy can be calculated from the Hugoniot equations. The temperature T can then be calculated by assuming a particular equation of state. The direct measurement of temperature during shock compression would make it possible to refine and extend the present knowledge of equations of state.

Measurement of temperatures produced by shock waves has been attempted by only a few workers. These measurements have been of two types: (1) measurement of residual temperatures and (2) measurement of temperature achieved during shock compression.

For opaque materials, it has not yet been possible to measure the temperature achieved during the shock compression although several workers have attempted to use thermocouples to do so (see section 6.1.3). The closest agreement between experiment and theory was the work of Buzhinskii and Samylov (1970) whose results differed by 60 per cent from calculated temperatures. Taylor (1963) made measurements of the residual temperatures of copper shocked to stresses ranging from 900 to 1700 kbar. These measurements were in good agreement with published temperature calculations (McQueen and Marsh, 1960). Taylor measured the total radiation per unit area visible to a photomultiplier tube, and a vacuum furnace was used to calibrate the photomultiplier tube.

In transparent materials, the temperature of the shocked sample can be determined from the brightness of the radiation emitted during the shock compression. At a high enough stress level, the originally transparent material becomes opaque and the light emission from the shock front is observed through a layer of still uncompressed material.

Zel'dovich *et al.* (1958) determined the shock temperature in Plexiglas by observing photographically the light emission from the shock front in the blue (4020 Å) and red (6000 Å) regions of the spectrum. These authors apparently calibrate by photographing a source of known temperature (although no mention of calibration is made in the article referenced). Kormer *et al.* (1965) measured the temperature of shock-compressed NaCl and KCl up to 800 kbar. They used optical methods and compared the light emission from the shocked samples with that of a standard light source. They used photomultiplier tubes together with interference filters whose transmission maxima corresponded to 4780 and 6250 Å. In this study they were able to determine the melting curves for NaCl and KCl from 540 to 700 kbar and from 330 to 480 kbar, respectively.

Some caution should be exercised in determining the equilibrium shock temperature from the emitted radiation. In certain temperature ranges, the observed temperature is influenced strongly by a layer of electrons just behind the shock front. The nature of this influence is discussed in more detail in section 6.2.

5.6   EXPERIMENTAL DETERMINATION OF CONSTITUTIVE RELATIONS

In most experiments, shock velocity and one other parameter are measured. If only initial and final states behind a stable shock are measured, the other parameters are calculated through use of the Hugoniot jump conditions (Rice *et al.*, 1958; Duvall and Fowles, 1963). Increasingly, however, measurements are being made as a function of time, and intermediate or nonsteady states are determined. Details of the material response during shock compression and subsequent release are recorded in many of these experiments. An especially interesting situation occurs when in-material gauges are used to record stress or particle velocity, or both, versus time at two or more planes in the specimen. If the records are carefully correlated in time, constitutive relations among stress, wave velocity, particle velocity, and density may be calculated without further restrictive assumptions (Fowles and Williams, 1970). Cowperthwaite and Williams (1971) show that either a pair of stress gauges or a pair of particle velocity gauges are sufficient to determine constitutive relations in steady, simple isentropic, and simple nonisentropic flows, but for general flows, constitutive relations among stress, particle velocity, and density can be calculated with three stress gauges but not with three particle-velocity gauges. We shall now briefly review the principles behind the analysis of such in-material gauge data.

We first recall from equations (68) and (69) in section 4 that

$$(U_\sigma - u) = \frac{\rho_0}{\rho} \left(\frac{\partial h}{\partial t} \right)_\sigma,$$

and

$$(U_u - u) = \frac{\rho_0}{\rho} \left(\frac{\partial h}{\partial t} \right)_u,$$

where we also recall that h is the original position of a particle at $t = 0$, and U_σ and U_u are the velocities in laboratory coordinates of waves of constant stress and particle velocity, respectively. Thus, written in Lagrange coordinates, equations (53) and (54) in section 4 become, respectively,

$$u(h,\, t) - u_0 = V_0 \int_{\sigma_0}^{\sigma\,(h,\, t)} \frac{\mathrm{d}\sigma}{(\partial h/\partial t)_\sigma} \tag{107}$$

and

$$\frac{V(h,\, t)}{V_0} = 1 - \int_{u_0}^{u\,(h,\, t)} \frac{\mathrm{d}u}{(\partial h/\partial t)_u}. \tag{108}$$

Equation (108) is used with particle velocity gauges and equation (107) is used with stress gauges.

In determining constitutive relations with multiple Lagrange gauges, the difference between the quantities $(\partial h/\partial t)_\sigma$ and $(\partial h/\partial t)_u$ may be very important. We will use the notation of Fowles (Fowles and Williams, 1970) and label these two quantities C_σ and C_u respectively. In Fig. 43,

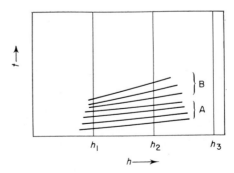

Fig. 43. Position–time diagram for two Lagrangian gauges. The lines labelled A and B are lines of either constant stress or particle velocity.

an h–t diagram is shown for two Lagrange gauges, either stress or particle velocity. The derivatives C_σ and C_u are, in practice, evaluated by the approximations

$$C_\sigma = \frac{h_2 - h_1}{t_2(\sigma) - t_1(\sigma)} \tag{109}$$

and

$$C_u = \frac{h_2 - h_1}{t_2(u) - t_1(u)}, \tag{110}$$

respectively. In general, $C_\sigma \neq C_u$ and equations (109) and (110) are only approximations to these derivatives. Therefore constitutive relations calculated from equations (107) or (108), or both, are valid only to the extent that equations (109) and (110) are valid. Cowperthwaite and Williams (1971) showed that exact constitutive relationships between σ, u, and V can be determined if the waves are steady state, simple isentropic, or simple nonisentropic. For these three types of flow, these authors showed that $C_\sigma = C_u$, and the C's are straight lines in an h–t diagram.

To determine whether a steady-state or simple wave obtains in a particular experiment, three Lagrange gauges should be used. With the three σ–t or u–t profiles obtained, the validity of equations (109) or (110) can be checked. Points of constant stress (σ) or particle velocity (u) are plotted for each gauge in an h–t diagram (see Fig. 43). If the lines joining points of constant σ or u are straight and parallel (as in region A of Fig. 43), then the wave is a steady-state one. For steady-state waves, equations (107) and (108) become the well-known Rankine–Hugoniot equations for the conservation of mass and momentum (see Fowles and Williams, 1970). If the lines joining points of constant σ or u are straight but not parallel (as in region B of Fig. 43), then the wave is a simple one. Equations (107) and (108) are then used to compute relations between stress, particle velocity, and specific volume. If the lines joining points of constant σ or u are neither straight nor parallel, then the wave is neither steady state nor simple, and $C_\sigma \neq C_u$. However, within the limits of the finite difference approximations, exact constitutive relations between some of the variables can still be calculated between pairs of gauges using either equation (107) or (108). With three gauges, σ–u relations can be calculated, and with three particle velocity gauges, u–V relations can be calculated (see Cowperthwaite and Williams, 1971).

For more general types of waves with $C_\sigma \neq C_u$, a better estimate of the derivatives can be obtained from a parabolic fit of the curves of constant σ or u (Cowperthwaite and Williams, 1971). In this approximation the expression for C_σ and C_u in the interval $h_1 \leq h \leq h_3$, $t_1 \leq t \leq t_3$ is

$$C_{\sigma, u} = \left(\frac{h_2 - h_1}{t_2 - t_1}\right) + \left[\left(\frac{h_2 - h_1}{t_2 - t_1}\right) - \left(\frac{h_3 - h_1}{t_3 - t_1}\right)\right]\left(\frac{t_2 + t_1 - 2t}{t_3 - t_2}\right). \quad (111)$$

With $C_\sigma \neq C_u$, relations between stress, particle velocity, and specific volume can be calculated in a region covered by three stress gauges but not in one covered by three particle-velocity gauges. Specifically three experimental σ–t profiles can be used with equations (107) and (111) to obtain u–t profiles. These calculated u–t profiles can then be used with equations (108) and (111) to obtain V–t profiles, thus providing the necessary relations between σ, u, and V. Note that errors made in calculating u–t profiles are compounded in calculating the V–t profiles. Three experimental u–t profiles can be used with equations (108) and (111) to obtain V–t profiles, but σ–t profiles cannot be obtained since C_σ cannot be calculated without additional assumptions or data. In practice, a combination stress–particle-velocity gauge (see below) can be used at one of the gauge positions to provide σ–t data. In this case, constitutive relations between σ, u, and V can be determined with two particle-velocity gauges and a combination stress–particle-velocity gauge. It is thus seen that the stress gauge is inherently a more powerful tool than the particle-velocity gauge.

The above analysis with multiple embedded Lagrange gauges is applicable to release waves as well as to compression waves.

The discussion above is applicable to one-dimensional nondecaying stress waves only. The analysis of data from in-material gauges for one-dimensional decaying waves has been discussed by Murri (1972) and Fowles (1973). Cowperthwaite (in Keough et al., 1971), Fowles (1970), and Grady (1973) have extended the analysis of data from Lagrangian gauges to cases of spherical flow for which the stress wave is of course attentuating with distance and time.

Cowperthwaite (1972) has also developed a method for modelling the energy release rate in condensed explosives using multiple Lagrange gauges in experimental studies of initiation and propagation of detonation.

Recently Chen et al. (1972) have developed a theory for the analysis of unsteady waves that is based on acceleration wave theory. In this theory, the experimental wave profiles are approximated by a sequence of chords. The discontinuities of slope where these chords join are assumed to be acceleration waves. The velocity of the acceleration wave is a critical parameter that appears in the governing equations for the data analysis, and it must be known or measured to carry out the analysis. However, this instantaneous sound speed cannot be measured experimentally. The approximation of this wave speed is the same as C_u or C_σ above, which, if known or measured, can be used to determine constitutive relations among stress, volume, and particle velocity by the more exact analysis method previously outlined above.

The use of multiple in-material gauges requires precise time correlation between gauge records. This can be achieved by imposing a common Z-axis modulation signal on all oscillograms. Gauges should be behind or almost behind each other so that the same signal is monitored at successive times at each gauge. The gauges need to be constructed and placed in the material in such a way that perturbation of the flow is insignificant. This can be achieved by making the total gauge package thin and close to the shock impedance of the specimen material.

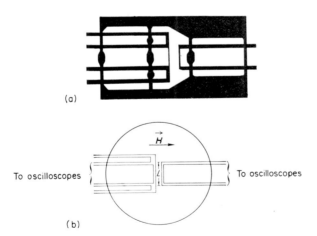

(a)

To oscilloscopes

$\vec{H}$

To oscilloscopes

(b)

Fig. 44. Experimental arrangement of combined stress and particle-velocity gauges. Target is mounted between poles of a magnet. In general, the effective length of the gauge element, to be used in calculating the voltage produced by particle motion, will be less than the length l shown above. (a) Photoetched gauges before mounting. After mounting, the supporting structures are cut away leaving gauges as shown in (b). (b) Front view of a stress gauge and a particle velocity gauge at one plane of the target.

Another promising development has been the use of a combination gauge to measure both stress and particle velocity simultaneously at the same plane (Murri *et al.*, 1969). In this technique an in-material manganin stress gauge and an in-material particle velocity gauge are used together (Fig. 44). The active elements *l* of both gauges are parallel and of the same dimensions. The leads are parallel to the magnetic

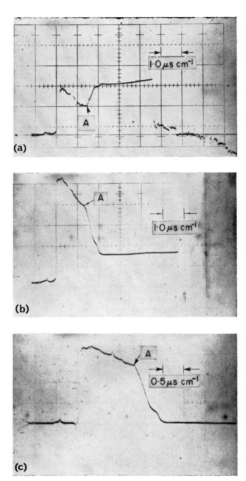

Fig. 45. Gauge records using combined gauges and in-contact explosive. Records show the relief wave following behind peak stress and the relief wave originating from the free surface beginning at point A. (a) Velocity gauge record. (b) Stress gauge record (containing a particle velocity component). (c) Corrected stress gauge record.

field. The particle velocity gauge works as described above in section 5.2.2 (Fig. 35). The output from the stress gauge includes the usual stress–time measurement as described above in section 5.2.2 but in addition it includes a particle velocity contribution because of its motion in the magnetic field. Since the two active elements have the same dimensions, each gauge records the same particle-velocity–time profile. Therefore, if the output from the particle velocity gauge is subtracted from the output from the manganin gauge, the result is the stress–time profile (Fig. 45). The stress gauge may also be oriented so that the active element is parallel to the direction of the magnetic field with the leads existing either parallel or perpendicular to the field direction. In this orientation, no voltage signal will be produced as a result of the stress gauge motion in the magnetic field and the gauge output will be proportional to stress only.

6 Effects of shock waves in solids

In this section, we shall discuss the effects produced by the passage of stress waves through materials. These effects fall rather naturally into two categories, dynamic effects and residual effects. Dynamic effects are those that are measured or observed during the time the material is under stress. Residual effects are those that are measured or observed in the material after it has been stressed and relieved. The dynamic effects will be considered first, followed by a discussion of residual effects.

The review articles by Doran and Linde (1966) and Hamann (1966) provide a rather comprehensive review of the literature on shock-produced effects in solids through December 1965. Al'tshuler (1965) and Dremin and Breusov (1968) have published excellent reviews on the use of shock waves in high-pressure physics and chemistry. Therefore we will concentrate on those papers published since these reviews.

6.1 ELECTRICAL EFFECTS

6.1.1 Conductivity

The effect of static pressure on the electrical conductivity of materials has been studied for many years, particularly by Bridgman (1952), and also by Swenson (1960). From these studies, it has been found that pressure can decrease or completely eliminate the energy gap between valence and conduction bands in solids with the result that many semi-conducting materials become metallic under pressure. In some cases, InSb for example (Darnell and Libby, 1963; Hanneman et al., 1964), the metallic phase can be retained at atmospheric pressure by thermally quenching the specimen before the pressure is released. In metals, a variety of behaviour is present, including phase changes resulting from rearrangement of electron configurations (Kohn, 1965).

When a material is shocked, the temperature increases as well as the stress. Not only is the specimen compressed, but large shear stresses can also be produced in the shock front (see sections 3 and 4.2.2). Thus the net effect of shock loading on the electrical conductivity can be a combination of lattice compression and distortion as well as a rise in temperature.

The review articles by Duvall and Fowles (1963), Hamann (1966), Doran and Linde (1966), especially Styris and Duvall (1970), and Keeler and Royce (1971, pp. 106–125), deal in some detail with measurements of electrical conductivity during shock loading. Therefore in this section we shall limit our discussion to more recent results. These will be grouped according to whether the material studied is normally a metal, semiconductor, or insulator.

a. *Metals* Iron has received considerable interest in both static and dynamic experiments, probably because of the phase change occurring at about 130 kbar. This transition was observed first dynamically (Bancroft et al., 1956) and subsequently statically (Balchan and Drickamer, 1961b). Electrical resistance measurements were made on iron during shock loading by Fuller and Price (1962), Wong et al. (1968), Keeler and Mitchell (1969), and Wong (1969) up to stresses of the order of 350 kbar.

The dynamic resistance measurements were made on samples in the form of wires or thin foil strips enclosed in an insulator. A constant current is passed through the specimens, and their resistance is monitored by oscilloscopes. In iron, there is a sudden increase in resistance beginning at about 120 kbar (Fuller and Price, 1962; Wong et al. 1968). This resistance change is the result of the phase change in iron. The data of Wong et al. differ somewhat from those of Fuller and Price in the sharpness of the resistance rise, although this discrepancy is probably due to a difference in impurities and perhaps to contact resistance (Fuller and Price used a two-terminal measuring system, whereas Wong et al. used a four-terminal one).

Wong (1969) and Keeler and Mitchell (1969) observed eddy-current spikes in the resistance–time records of shock-loaded iron. Wong also observed these spikes in nickel. When the constant current, used to observe the resistance as previously described, passes through a magnetic material, a sizeable magnetic field is produced in the magnetic material. Typically the viewing currents are of the order of 3 A and the resulting magnetic fields of the order of 100 Oe. Eddy currents are therefore produced in the specimen when a partial or complete demagnetization occurs during shock loading. These eddy currents appear as voltage spikes in the observed resistance-time record. Wong (1969) suggested three mechanisms that can cause partial or total demagnetization during shock compression. They are (1) shock-induced phase change from a magnetic phase to a nonmagnetic phase, such as the transition from α-iron to ϵ-iron; (2) rotation of the magnetic domains arising from anisotropic elastic strain associated with the shock (see Royce, 1966); (3) nonequilibrium spin fluctuations caused by overlapping electron clouds (3d-electrons in the case of iron) or neighbouring magnetic atoms as a result of shock compression. In the same paper Wong suggested a method by which the stress limits of magnetic phase transitions can be determined using these eddy-current spikes and double shocking the specimen.

Manganin (an alloy of copper, manganese, and nickel with an approximate composition of 80 to 85 per cent Cu, 10 to 15 per cent Mn, and 2 to 5 per cent Ni) has been extensively studied because of its usefulness as a stress transducer material. The results of the work performed on this material were discussed in section 5.1.2. We refer the reader to that section for these results and experimental details.

Several other metals have been studied to assess their value as stress transducer materials. The piezoresistance of calcium, lithium, cadmium, indium, lead, bismuth, and ytterbium was discussed in section 5.1.2.

b. *Semiconductors* Extensive literature exists on the effect of pressure on the properties of semiconductors. An excellent review was given by Paul and Warshauer (1963), and more recent results were discussed by Falicov (1965), Fritzsche (1965), and Adler and Brooks (1965). The chief effect of the pressure (static) is the reduction of the energy gap between the valence and conduction bands, although the shape of these bands can also change.

Compression from stress waves is also expected to reduce the energy gap and to heat the compressed material because of the nonisentropic nature of stress-wave compression. Hamann (1966) reviewed the work performed up to about 1965. Since that time, not much has been done on the conductivity of shock-compressed semiconductors, except for some work on silicon (Coleburn et al., 1972) which is discussed in section 6.1.2(e).

Graham et al. (1965b, 1966) reported on work done on germanium shock-loaded by impact along the [111] crystallographic direction. These authors measured the resistivity versus time up to the peak of the elastic wave $(44 \pm 4 \text{ kbar})$. They assumed that up to 44 kbar at least, the resistivity of germanium behaves intrinsically for large shear strain. With this assumption and a slight temperature correction, they were able to calculate the change in the energy gap produced by shear strain, since the energy gap change induced by volumetric compression had been determined from hydrostatic experiments. Their value of -9.0 eV per unit strain for the shear strain contribution in the [111] direction is in good agreement (both sign and magnitude) with that predicted from theory for silicon (see Goroff and Kleinman, 1963).

c. *Insulators* The early work on the shock-induced conductivity of insulating material has been reviewed by Doran and Linde (1966), Hamann (1966), and Duvall and Fowles (1963). Most of the research has been done on the alkali halides, perhaps because of the available data on all aspects of these ionic solids.

The conduction mechanism in alkali halides is somewhat uncertain. Originally Alder and Christian (1956) interpreted the increased conductivity as a transition to a metallic state. Doran and Murri (1964) questioned this interpretation and suggested that the data could equally well be explained on the basis of increased ionic conductivity. Al'tshuler et al. (1961) studied NaCl and interpreted their results as a temperature dependence of ionic conductivity. The most recent results (for NaCl) are those of Kormer et al. (1966a) who proposed that the conductivity is electronic in nature. In their model (based on optical absorption measurements, see section 6.2,) the alkali halide is transformed by the shock compression into a semiconductor with donor levels produced by the plastic flow at the shock front. The accompanying temperature rise then thermally excites electrons from the donor levels to the conduction band. Typical values of the conductivity, free-electron mobility, and concentration for $\sigma = 465$ kbar are: $1.2 \times 10^{-1} \ \Omega^{-1} \ \text{cm}^{-1}$, 2.31 $\text{cm}^2 \ \text{s}^{-1} \ \text{V}^{-1}$, and $3.2 \times 10^{17} \ \text{cm}^{-3}$, respectively.

Ahrens (1966) studied MgO single crystals shocked in the [100] direction to about 920 kbar. He observed a marked increase in the conductivity from 10^{-5} to 10^{-3} Ω^{-1} cm^{-1} for stresses of the order of 920 kbar; below this stress no increase was detectable. The conductivity mechanism was not determined but the author suggests that the conduction may be the result of electronic processes or ionic transport.

Kuleshova (1969) studied the increase in conductivity of shock-compressed boron nitride, potassium chloride, and polytetrafluoroethylene. Measurements were made on BN and single crystal KCl in the stress range 130 to 500 kbar and on polytetrafluoroethylene at stresses of 300 to 800 kbar. Whereas most investigators measured the conductivity in the direction of shock-wave propagation, Kuleshova measured it perpendicular to the direction of shock propagation. In this way, he was able to follow the development of the conductivity increase, because as the shock wave traverses the sample, more and more of the sample becomes conducting. (Murri and Doran (1965) noted an order of magnitude increase in the final conductivity of CsI in the direction transverse to the direction of shock propagation.) For BN, Kuleshova found that the logarithm of the conductivity was a linear function of temperature for stresses between 300 and 800 kbars with an activation energy of 3.0 eV. For KCl, the conductivity was determined in the solid and the melt. In the melt (330 to 480 kbar), the conductivity was a constant 1.45 Ω^{-1} cm^{-1}. In the solid, the conductivity was high for a relatively short (0.3 to 0.5 μs) period, after which it decreased to a value less than 10^{-3} Ω^{-1} cm^{-1}. Kuleshova suggests that the decrease in conductivity is caused by a recombination of electrons and holes which reduces the concentration of charge carriers. A sudden increase in the conductivity of polytetrafluoroethylene was observed at stresses of 300 to 800 kbar, but this increase occurred only after a time interval $\Delta\tau$ following the passage of the shock wave. This time interval depended on the peak stress and also coincided with the arrival of relief waves at the electrical contacts. Kuleshova suggests that the isentropic expansion increases the relative importance of the temperature causing the polytetrafluoroethylene to break down. An alternative explanation (although Kuleshova does not discuss it) is that the polytetrafluoroethylene melts on unloading with a sudden increase in conductivity at melting.

The melting hypothesis is substantiated by preliminary data of Champion and Bartell (1970) and Champion (1972) for high- and low-density polyethylene and polytetrafluoroethylene. They find the conductivity behind the shock wave to be time dependent and to increase upon unloading by 2 to 3 orders of magnitude.

6.1.2   *Electrical polarization and depolarization*

The term "shock polarization" is used to describe the production of charge on opposite faces of the dielectric material as the shock wave travels from one face to the other. For piezoelectric materials, such a phenomenon would be expected. However, shock polarization has been observed in several materials that are not piezoelectric, notably plastics and alkali halide crystals.

a. *Quartz*   The shock polarization of quartz has been studied extensively by Graham *et al.* (1965a) and reviewed by Doran and Linde (1966). Much of the interest in quartz has resulted from its usefulness as a stress transducer (section 5.1). For a discussion of experimental techniques for studying piezoelectric properties under conditions of transient stress, see Graham (1961) and Graham *et al.* (1965a). Graham (1962) reported that dielectric breakdown occurs if the quartz is oriented so that the polarization is opposite the direction of shock propagation. Graham and Halpin (1968) extended this initial study of dielectric breakdown and showed that up to at least 30 kbar breakdown occurs only when the directions of shock propagation and polarization are opposite. Their measurements showed that breakdown occurs at a stress greater than 10 but less than 13 kbar. When breakdown occurs, the current in the external short circuit connecting the faces of the shock-loaded quartz is characterized by an initial jump as the shock wave enters the quartz, followed by a sharp drop. The sharp drop in external current is a result of the dielectric breakdown caused by the high internal field. During the shock transit time in the quartz, the internal field is reduced and the quartz shows a tendency to recover from breakdown to essentially infinite resistivity.

Graham and Halpin also formulated a mathematical model, including finite resistivity, that permitted solutions for the internal electrical fields and resistivity in terms of the measured current. Computations with this model predicted that the electric field in the stressed portion of the quartz is about 7.0×10^5 V cm^{-1} at breakdown (an order of magnitude less than the observed value at atmospheric pressure). Their computations also predicted that the quartz recovers from breakdown when the electric field is quenched to about 1.9×10^5 V cm^{-1}.

This critical field is independent of the stress, at least between 13 and 36 kbar (the highest stress in these experiments). These authors suggest that breakdown is initiated from a source of free electrons at the shock wavefront. The electrons most probably result from dislocation motion in the wavefront.

Jones (1967) studied the piezoelectric response of quartz at 79 K. He found only a slight change in the piezoelectric factor at low temperatures. This factor depends on stress and relates the observed current and the stress in uniaxial strain. At 298 K the value is $(2.01 \times 10^{-8} + 1.1 \times 10^{-10} \sigma_x)$ (C cm^{-2} kbar^{-1}) (Graham et al., 1965a), and at 79 K Jones reported a value of $(2.11 \times 10^{-8} + 5.5 \times 10^{-11} \sigma_x)$ (C cm^{-2} kbar^{-1}) for stresses between 6 and 21 kbar, where σ_x is the uniaxial stress in kbar. Rohde and Jones (1968) made measurements at 573 K in the stress range of 6 to 21 kbar. Their results indicate a decrease in the piezoelectric factor to $(1.895 \times 10^{-8} + 1.470 \times 10^{-10} \sigma_x)$ (C cm^{-2} kbar^{-1}). They also observed current relaxation above 14 kbar.

Graham (1972b) has utilized experimental measurements of the piezoelectric current in X-cut-quartz to determine the finite-strain piezoelectric and elastic constitutive relations for uniaxial strain in quartz. The method of analysis seems generally applicable to other piezoelectric materials which have large elastic limits under shock loading. He has also developed a general method for calculating the third and fourth order longitudinal elastic constants and has applied the method to shock compression data for sapphire and fused quartz (Graham, 1972c). The method is based on the extension of finite strain theory to conditions of shock compression by Fowles (see Fowles, 1967).

b. *Ferroelectrics*   Doran and Linde (1966) and Keeler and Royce (1971, pp. 126–138) have discussed the shock effects produced in feroelectrics; this class of materials will not be treated here except to mention new results published since their reviews. Instead of causing polarization, the passage of a shock wave may destroy the dielectric polarization; for example, it may depole an already poled ferroelectric. A poled ferroelectric material subjected to a stress wave produces a current, under short-circuit conditions, by a depoling process in the stressed region. On the other hand, piezoelectric materials that are not ferroelectric, e.g. quartz (Graham et al., 1965a), generate a current by polarization in the stressed region.

Cutchen (1966) observed a polarity effect similar to that observed by Graham (1962) in quartz in dynamically stressed ferroelectric lead zirconate titanate ceramic (PZT), i.e. the voltage produced depends upon the direction in which the stress wave is travelling with respect to the polarization direction. This polarity effect had not been observed in ferroelectrics before, possibly because of the lower quality of ceramic previously available. The higher quality has resulted in a more homogeneous microstructure, greater density, and a higher static dielectric strength. Cutchen concluded that p-type mobile charge carriers are being liberated at the stress front and are in turn acted upon by the local internal fields. The charge liberation is postulated to be the result of ionization and failure mechanisms. Graham (1962) reported a similar effect in alpha quartz in which electrons are liberated. The action of the local fields on the liberated charge is found to result in dielectric degradation when the stress wave travels in a direction opposite to that of polarization.

Halpin (1968) reported a quantitative analysis of short-circuit current waveshapes from shocked ferroelectric materials that includes the effect of conduction. He applied this model to previous results (Halpin, 1966) and obtained estimates of the resistivities of shock-loaded ferroelectrics. Halpin's model predicts a resistivity of the order of 100 Ω cm with accompanying electric fields greater than about 20 000 V cm^{-1} for ferroelectrics shocked to between 25 and 35 kbar.

Doran (1968) studied the shock compression of barium titanate and 95/5 lead zirconate titanate in the ranges 5 to 200 kbar and 2 to 140 kbar, respectively. The specimens used by Doran were all unpoled.

Lysne (1973) has studied the dielectric breakdown of shock-loaded 65/35 PZT as a function of stress and electric field strength. He concluded that dielectric breakdown was not explicitly a function of stress up to at least 24 kbar. The breakdown was not instantaneous and for electric fields greater than the breakdown threshold the time between loading and breakdown decreases with increasing electric fields. He also observes that the voltage response of a partially poled specimen may be greater at high stresses than that for a fully poled specimen shocked to the same stress.

Novitskii et al. (1973) have also investigated the depolarization of $BaTiO_3$, $PbTiO_3$, and $LiNbO_3$ during shock loading.

c. *Alkali halides* Eichelberger and Hauver (1962) were the first to report the observation of shock-produced polarization in materials that are not naturally piezoelectric. Allison (1965) developed a phenomenological theory that assumes a shock-induced change in the dielectric constant and decay of the polarization signal with a characteristic relaxation time τ. Hauver (1965) was successful in using this theory to explain his data for Plexiglas and polystyrene. These early results were reviewed by Doran and Linde (1966) and Hamann (1966).

The earliest published observations of shock polarization in alkali halide crystals are those of Ivanov et al. (1965), Ahrens (1966), and Linde et al. (1966), which report results for stresses from 10 to 500 kbar. Since these first results, several papers have been published, principally by Russian authors.

Zel'dovich (1968) calculated the emf in a short-circuit solid-dielectric capacitor subjected to shock compression. He assumed that no changes in dielectric constant occur, that the dielectric is not compressed, and that material behind the shock front is conducting while that ahead of the front is not. He further proposed that decay of the polarization signal is due to conductivity relaxation with a characteristic time θ. This relaxation results from the compensation of the oriented dipoles (i.e. bound surface or volume charges) by free charge carriers generated by the shock. Since no account is taken of changes in dielectric constant or of compression, the range of applicability of the theory is greatly reduced.

Yakushev et al. (1968) made direct measurements of the polarization relaxation time τ (see Allison, 1965; Hauver, 1965) in Plexiglas in the stress range 100 to 170 kbar. They found τ to be independent of stress in that range and equal to 1.1 $\mu s \pm 40$ per cent. This result agrees well with that of Hauver (1965). Their direct measurement was accomplished by excluding any shock reflection from the second voltage measuring electrode. They were able to do this by using a thin aluminium foil backed by additional dielectric as the second measuring electrode.

Ivanov et al. (1968a) shocked NaCl in the [100], [110], and [111] directions as well as pressed polycrystalline samples. These authors applied the theories of Allison (1965) and Zel'dovich (1968) to their data and found them applicable only in some special cases. Ivanov et al. (1968b) developed a theory that combined the polarization relaxation mechanisms of Allison (thermal disorientation) and Zel'dovich (conductivity behind the shock wave) and in addition allowed for compression and change in the dielectric constant. Their quantitative solutions fit the available data for several particular cases.

Mineev et al. (1968) investigated shock polarization in single crystals of LiF, KCl, and KBr at stresses between 20 and 200 kbar and up to

6

1000 kbar for LiF. All crystals were shocked in the [100] direction. They observed a sharp change in the polarity and magnitude of the polarization voltage in KBr and KCl similar to that previously observed in NaCl (Ivanov et al., 1965; Linde et al., 1966; Ivanov et al., 1968a). These authors discuss the combined results of their work and those of previous studies and conclude that shock polarization is the result of crystal lattice imperfections produced during plastic deformation by the shock wave. Specifically they suggest that the polarization signals are the result of two simultaneously occurring processes: (1) the production of cation dipoles, i.e. a cation vacancy caused by the displacement of cations from lattice sites; (2) the orientation of these dipoles in the direction of the shock-wave propagation. They also suggest that up to compression of about $\rho/\rho_0 \approx 1.3$, the ion dimensions influence the polarization signal.

Tyunyaev et al. (1969a) studied the role of ion mass in the shock polarization of ionic single crystals. They compared the results for Li^7H and Li^6D and for RbI and KI and concluded that ion mass does not influence the polarization current. These results corroborated the findings of Linde et al. (1966). However, Tyunyaev et al. made three general observations on the basis of analysis of available data for 10 ionic crystals. Given the same compression, and crystal orientation relative to the shock wave, then: (1) for compounds with the NaCl lattice and for a given cation, the value of the polarization remains constant when the anion is replaced; (2) in the same compound, the polarization decreases with increasing cation radius; (3) regardless of the lattice type, the polarization increases with increasing dielectric constant. Observations (1) and (2) suggest that the cation dimension is a controlling factor in determining the value of the shock polarization.

Tyunyaev et al. (1969b) carried out impact loading experiments on single crystal CsI, both pure and doped, to study the role of dislocations on the impact polarization of alkali halides. Slip in CsI type lattices takes place on {110} planes in the ⟨100⟩ directions; hence slip should occur rather easily for compression in the [110] or [111] directions. If dislocation motion is a factor in the polarization phenomena, a smaller polarization signal should result for compression in the [100] direction than in the [110] or [111] directions. Following this reasoning, Tyunyaev et al. shocked CsI in these three directions. In addition they shocked two types of doped CsI, one doped with Pb and the other with Ba. The addition of Ba strengthens CsI, whereas the addition of Pb shows no effect; thus, the polarization signal for a Ba doped crystal should be reduced if dislocation motion is important. Based on their results, Tyunyaev *et al.* concluded the following: (1) for low deformations ($\rho/\rho_0 \lesssim 1.25$) dislocation motion does not affect the shock polarization; (2) at larger compression (up to $\rho/\rho_0 \sim 1.4$) dislocations may be influential. However, these authors feel that other point defects may contribute, and indeed several processes may be taking place simultaneously, so that the influence of some given single process may predominate at certain stages of the compression.

The effect of impurities on the polarization signal has been studied further by Tyunyaev *et al.* (1970). These workers used NaCl crystals containing controlled amounts of Ca^{2+} and Li^+. They found that the addition of divalent Ca^{2+} changed the sign and value of the polarization signal. The effect of Ca^{2+} also depended upon the concentration. However, monovalent Li^+, for the same concentration, caused no change in the polarization signal. The authors of this paper suggested that the experimental observations could be understood in terms of nondislocation mechanisms such as reorientation of charged complexes or the shifting of a Cl^+ ion to a cation vacancy (see also Mineev *et al.*, 1968; Tyunyaev *et al.*, 1969b).

Tyunyaev and Mineev (1971) have studied shock-wave polarization in LiF as a function of heat treatment, crystallographic direction, and doping with Mg^{2+}. They found that quenching of pure LiF had no effect in the stress range investigated. However quenching of Mg^- and Mg^{2+} doped crystals caused the polarization signal to increase. The orientation dependence of the polarization signal did not correspond to the plasticity anisotropy.

Wong *et al.* (1969) carried out perhaps the most complete investigation of shock polarization. In their paper, only results for NaCl are discussed. They present a summary of previously published data for NaCl and in addition new results for crystals doped with Ca^{2+}, Ba^{2+}, Mg^{2+}, and OH^- impurity ions as well as for crystals with special mechanical and heat treatment to change the initial dislocation density. These investigators suggest a model for the polarization signal based on dislocation dynamics. Specifically, their model proposes that the charged dislocation cores in NaCl, both positive and negative, attract cation and anion vacancies producing a Debye–Hückel cloud (Eshelby *et al.*, 1958). As the stress wave moves through the crystal, the charged dislocation cores are separated from their vacancy clouds thus produc-

ing electric dipoles. This theory combines the idea of charged dislocation motion, used successfully to interpret quasi-static deformation
results (Whitworth, 1965), with that of point defects. This model
depends primarily on the initial Schottky defect concentration and
therefore should be essentially independent of changes in dislocation
density caused by heat treatment and mechanical working of the crystals before shock loading. Wong *et al.* showed this to be the case. In
addition, they were able to explain qualitatively all the existing data
for NaCl including the change of signal polarity at about 10 and 110 kbar
and the variations in signal profiles observed when the crystals are
shocked along different crystallographic directions.

Yakushev (1972) has discussed the effect of nonuniformity of the
shock wave on the polarization signal. A model is also proposed for
evaluating the polarization current in relation to the rise time of the
shock wave.

d. *Electrets* Champion (1968) noted that a fast-rising voltage signal was
produced when an anodized aluminium layer was compressed by a
shock wave. In a later publication (Champion, 1969), a more thorough
study was carried out. Measurements were made on anodized aluminium
layers 0.005-cm thick from 5 to 220 kbar. The signal polarity depended
upon whether or not the aluminium substrate on which the anodized
layer was formed was grounded. A linear relation between the peak
voltage and stress was observed up to 75 kbar in aluminium. At 150 kbar
there was considerable scatter in the data, and at 220 kbar the output
signal was nearly zero. Champion was able to interpret his results in
terms of an electret-like behaviour of the anodized layers. He developed
a macroscopic model based on this behaviour that predicted the major
features of the electrical signal. Ionic species incorporated into the
oxide layers during the process of anodizing are thought to produce the
permanent polarization observed in the layers. A static charge of
10^{-9} C cm^{-2}, which persists for at least three months, was measured
on the layers.

e. *Semiconductors* Mineev *et al.* (1967, 1971) reported polarization signals
from impact-loaded semiconductors. They presented results of a
study on p-type single crystal silicon shocked perpendicular to the
(111) plane as well as results from individual experiments with n-type
single crystals of silicon and germanium. These workers shocked the
p-type silicon to 200 kbar and made the following observations: (1) The
polarization current is positive and begins simultaneously with shock

entrance into the sample. (2) The polarization current is nearly constant during the shock transit time, except for the initial rise. (3) The voltage drop across the viewing resistor is nearly independent of this resistance between 93 and 1053 Ω. (4) The polarization current remains nearly constant as the sample thickness was varied from 0.1 to 0.6 cm. Mineev et al. proposed that the polarization mechanism is connected with impurities (their samples contained about 10^{17} impurity atoms per cm^{-3}) and that a double electric layer propagates with a constant velocity together with the front of the elastic wave. The results for n-type germanium and silicon indicated a polarization signal of the same polarity and approximately the same magnitude as for the p-type silicon.

Coleburn et al. (1972) have made electrical measurement in p-type silicon shock loaded in the (111) direction with stresses between 8 and 160 kbar. Their polarization results indicate only positive charges were produced during shockwave passage. The maximum signal occurred at a stress just below the HEL of 55 kbar. From electrical resistance measurements, they conclude that the doped silicon transforms to a metallic state at stresses near the HEL. At 133 kbar, the relative resistance increases significantly and indicates a polymorphic transition.

Mineev et al. (1972) investigated the emf observed upon shock loading cerium, europium, and ytterbium. Their studies indicate that the emfs produced are due principally to volume redistribution of the charge carriers in the shock front.

Ivanov et al. (1968c) have observed polarization signals in shock-loaded bismuth at a stress of 260 kbar. They conclude that the charge carriers are negative, most probably electrons.

6.1.3 Thermoelectric effect

Thermocouples have been used by several workers in an attempt to measure temperatures of materials during the passage of shock waves. These experimental efforts, although failing to measure temperature, have led to the discovery of what has been termed an "anomalous thermoelectric effect". It was found that when thermocouple junctions in metals experience shock compression, they generate a much greater emf than would be expected from the calculated shock temperature and zero-pressure thermopowers. The correction for static compression of thermocouples is small (Hanneman and Strong, 1965) and is generally in the wrong direction.

Jacquesson (1962) was the first to note this effect, although it was studied at about the same time in several laboratories (Plyukhin and Kologrivov, 1962; Doran and Ahrens, 1963; Palmer and Turner, 1964; and Crosnier et al., 1965). To illustrate the magnitude of this "anomalous" effect, we quote from the results of Crosnier et al. (1965). For a copper–constantan junction shocked to 400 kbar, the observed voltage is 73 mV. (See section 5, Fig. 29, for experimental arrangement.) The calculated shock temperature is 311°C, and a static copper–constantan thermocouple produced a signal of only 11.8 mV (signal corrected for pressure by $-3°C$).

Crosnier et al. (1965) performed many experiments with the bimetallic junctions in various geometries and combinations. They concluded: (1) The effect is similar to the normal thermoelectric effect except that pressure (or stress) is substituted for temperature (i.e. the notion of a hot junction is replaced by a shocked or compressed junction). (2) A bimetallic junction is necessary, since no signal is obtained from a copper–copper junction. (3) The junctions behave as voltage sources, with an internal resistance of the order of a fraction of an ohm, and not as current sources. (4) In the pressure range from 180 to about 1600 kbar the voltage versus pressure follows a linear empirical law given by

$$V(\text{millivolts}) = 0.45 \ \sigma(\text{kbar}) + 11.$$

(5) The rise time of the signal seems to be of the order of the shock rise time at the junction. The use of the effect as a stress transducer was discussed in section 5.1.4.

Jacquesson et al. (1970) extended the study of shocked bimetallic junctions to semiconductors. They studied copper–semiconductor junctions with intrinsic n- and p-type semiconductors using polycrystalline and single crystal specimens. They found that the signal polarity from the shocked junction is the same as the polarity of the dominant charge carrier in the semiconductor. Additionally, the signal magnitude was not greatly pressure dependent.

The chronometric property of the shocked junctions was used by Jacquesson et al. (1970) to study the shock-wave structure in germanium and bismuth. They observed the Hugoniot elastic limit in germanium and phase transitions in both germanium and bismuth.

Several attempts have been made to explain the mechanism of this effect. Plyukhin and Kologrivov (1962) suggested that the temperatures implied by the measurements were correct, although they were much higher than would be predicted. Crosnier *et al.* (1965), Conze *et al.* (1968), and Migault and Jaquesson (1968) interpreted the effect in terms of an electronic gas whose temperature exceeds that of the crystal lattice. Phonons are created by dislocation movement and lattice disturbances (shear and compression) and the phonon–electron interaction produces an electronic state in which the electron temperature is different from that of the crystal lattice.

The most comprehensive theoretical study was carried out by Migault (1970). Using linear transport theory, he was able to obtain the solution of two coupled Boltzman equations governing the distribution functions of electrons and phonons in a metal with a time-dependent stress distribution. With this solution, he was able to calculate the electric field acting on the electrons and hence the potential difference that exists across a bimetallic junction during shock compression. Migault's calculations agree in polarity and magnitude with many of the experimental results. His theory is restricted to linear effects, and it treats the materials involved as fluids. This assumption means that the motion of crystalline defects is neglected. At high pressures (above a hundred kbar or so), this approximation is not too serious a restriction. Any extension of the theory to include lower stresses will require that consideration be given to the strength of the materials and to the motion of defects.

A significant feature of Migault's theory is that the calculated potential difference contains an electrical effect in addition to the normal thermoelectric effect. Apparently, the effect of the shock wave on the lattice–electron system is to increase the phonon–electron interaction so that the current flow is increased and a larger voltage is observed across a bimetallic junction than would be expected from the temperature differential alone. A similar enhancement of the thermoelectric voltage occurs in metals and semiconductors at moderately low temperatures and is due to "phonon drag" effects (Herring, 1954; Geballe and Hull, 1955).

Buzhinskii and Samylov (1970) have made measurements of the temperature behind the shock wave in copper using a copper–nickel thermocouple. They report temperatures differing by only 60 per cent from predicted values. They suggest some of the discrepancy in the measurements of others may result from transient effects and small thicknesses of one of the metals used in the thermocouple junction.

6.1.4   Hall effect

Since measurement of the Hall effect during shock loading is closely related to the electrical conductivity measurements just described, we will discuss these results here. The only experimental data known to the authors are the preliminary data of Kennedy and Benedick (1967) and the more thorough results of Kennedy (1968).

These workers shock loaded doped n- and p-type silicon and germanium single crystals in the [100] crystallographic orientation. The input stress level in these experiments was 41 to 43 kbar for germanium and 35 kbar for silicon. In each case, this stress level was below the Hugoniot elastic limit. Of particular interest are the results of p-type, gallium doped [100] germanium and n-type [100] silicon with carrier concentrations of $\approx 10^{14}$ carriers per cm^3. In both these materials, the Hall voltage reversed polarity during the shock transit through the specimen.

The reason for this Hall voltage reversal is not known. Kennedy and Benedick point out, however, that the valence bands of both silicon and germanium consist of two overlapping bands so that two different effective masses for holes are possible. A theoretical model developed by Willardson et al. (1954) to calculate the behaviour of doped germanium as a function of energy band gap, temperature, and magnetic field predicts that a Hall voltage reversal may occur for small changes in temperature and energy band gap.

6.2   OPTICAL EFFECTS

A variety of optical phenomena occur during the passage of stress waves through materials. The refractive index is affected by both temperature and pressure, but the theoretical conclusion is that the compression will cause it to increase so long as ionization does not occur. In photoelastic materials, stress waves can produce optical anisotropy, which leads to double refraction. Changes in the optical absorption spectra of liquids and solids can also be produced by shock waves, and shock-induced changes in the opacity of materials have been observed. As the shock wave becomes stronger, the temperature increases and may become high enough to cause light emission. The band spectra of the molecules will be excited first (at about 1000 K) followed by the line spectra of the atoms as the temperature increases, and finally the continuous spectra will appear when ionization occurs. Careful measurements of this continuous emission have led to experimental determination of the temperature produced in a shock wave (see section 5.5).

Hamann (1966), Doran and Linde (1966), and Kormer (1968) published rather complete reviews of the optical phenomena listed above, and we will confine the following discussion to results that have appeared since their reviews.

Kormer *et al.* (1965) measured the light emission from the shock front in several alkali halides. From their measurements they were able to trace the melting curves of NaCl and KCl up to 700 kbar. These temperature measurements were described previously in section 5.5. They observed that the emission brightness increased with shock propagation distance in the sample until, starting at a certain thickness, the brightness became constant. This increase is connected with the increase in the layer of material compressed by the shock wave and with the opacity of this layer. The absorption coefficient of the compressed material is determined from the observed brightness-time dependence, the shock velocity, and the density of the material behind the shock front. The values obtained by Kormer *et al.* for NaCl were about 1.5 cm^{-1} at 465 kbar ($T \approx 2550$ K) and 10–12 cm^{-1} at 790 kbar ($T \approx 4850$ K).

In a later paper, Kormer *et al.* (1966a) discussed the physical nature of the light absorption. They suggested that as the shock wave propagates, dislocations are created in the material, leading to the formation of point defects on which electrons are localized. The temperature of the material in the shock wave is high enough to ionize these donor levels and thereby produce electrons in the conduction band. This is the same mechanism used to explain the results of electrical conductivity experiments (see section 6.1).

The range of temperatures measured by Kormer *et al.* (1965) (2500 K $< T <$ 5000 K) agrees very well with those calculated from basic shock theory, providing the influence of melting and the specific heat of the free electrons in the conduction band are included. However, for temperatures below 2500 K (and down to about 1000 K) and above 5000 K, a noticeable difference between theory and experiment is observed.

In the low-temperature region, Kirillov *et al.* (1968) found that the optically determined temperature exceeded that calculated from the Mie–Grüneisen equation of state. They showed that the observed radiation was not all thermal and was probably not related to electroluminescence. Electroluminescence is characterized by a pronounced wavelength dependence of the radiation intensity, and this dependence was not observed in their experiments. Therefore, they suggested that nonequilibrium between the electrons and the lattice exists just behind the shock front, and the observed emission is from these "hot" electrons. The mechanism for obtaining electrons in the conduction band is as discussed above. Kirillov *et al.* propose that the time to establish thermal equilibrium between the electrons and the lattice is $\tau \geqslant 10^{-7}$ s.

In the high-temperature region, the reverse is true; Kormer *et al.* (1969) found the optically measured temperature was less than the predicted one. These workers studied several alkali halides shocked to stresses from 500 to 5000 kbar. They also suggested that, at the temperatures produced ($\geqslant 5000$ K), a layer with nonequilibrium electron temperature exists just behind the shock front, and its emission determines the observed brightness temperature. The emission from deeper, hotter layers beind the shock front in the sample, where the electrons and the lattice are in equilibrium, is completely screened by this nonequilibrium layer.

The observed emission in both temperature regions is from electrons in the conduction band that are not yet in equilibrium with the lattice. In the low-temperature region, these electrons come from ionized donor levels produced by plastic deformation, and their temperature is higher than the equilibrium temperature. In the high-temperature region, a considerable transfer of electrons from the valence band to the conduction band takes place through multiple electron–phonon interaction (see Zel'dovich *et al.*, 1969). These electrons have not yet come to equilibrium with the lattice and form the layer of "cold" electrons that screens the radiation coming from the deeper equilibrium layers. Zel'dovich *et al.* (1969) discussed the kinetics of the electron excitation to the conduction band and developed a theory that accounts for the discrepancy between observed and calculated temperatures. Calculations of brightness temperatures based on their theory agree reasonably well with the experimental observations in both temperature regions. Perhaps it was through experimental luck that the agreement between theory and experiment was achieved, because Kormer *et al.* (1965)

unknowingly performed their experiments in the temperature range for which the kinetics of electron transfer into the conduction band are such that the observed and calculated temperatures agree.

Kormer *et al.* (1966b) studied the dependence of the refractive index on the density of the solid and liquid phases of ionic crystals during shock compression. They found that the refractive index changes relatively little with density as long as the shocked crystal remains in the solid phase, but the refractive index increases greatly when melting occurs during shock compression; the change in refractive index with density in the liquid phase is about 15 to 17 times larger than in the solid phase. The effect of crystallographic phase changes was also observed.

Yakusheva *et al.* (1970) made direct observations during shock compression of the absorptive capacity of single crystals containing F centres. Previous measurements of the formation of point defects and associated phenomena (Gager *et al.*, 1964; Doran and Linde, 1966; Linde and Doran, 1966) were made on crystals recovered after shock loading. These recovered crystals have not only been compressed by the shock wave but have also undergone plastic deformation in the unloading wave. Therefore, it is not clear whether lattice defects are generated in the shock front or in the rarefaction wave or both. Yakusheva *et al.* shock-loaded NaCl single crystals while simultaneously illuminating them with an explosive argon light source and recorded the reflected light on colour film with the aid of a streak camera. They found that pronounced bleaching of the coloured crystals occurred during the shock transit. The crystals were shocked at 30 and 180 kbar. The bleaching was attributed to the generation of a large number of electron traps in the shock front.

Gaffney and Ahrens (1972) and Gaffney (1972) have also measured the absorption spectra in solids during shock compression. In their system, a 60 000 K light beam enters the sample through the unshocked surface, passes through the shock front and is reflected by an internal mirror and leaves the sample by the reversed path. They have measured changes in the crystal-field spectra with stress in ruby and magnesium oxide.

6.3   PHASE TRANSITIONS

Some of the earliest shock wave work reported concerned observation of shock-induced, high-pressure phase transitions (Bancroft *et al.*, 1956;

Rice *et al.*, 1958). Since that time a great many materials have been observed to undergo phase transitions at certain stress ranges. The most favourable situation for observing the dynamic phase change is when the volume change on the $P-V$ Hugoniot is great enough for the cusp to give rise to a double (or triple) shock-wave structure over a certain pressure range, as explained in section 4.2.

Doran and Linde (1966) have given a recent review including most of the shock-induced phase transitions reported at the time. Other recent reviews are given by Hamann (1966) and Duvall and Fowles (1963).

Lysne *et al.* (1971) have written an excellent review of shock-induced phase transformation. Their review includes a comprehensive listing of known transformations, some description of experimental evidences of phase transformations, and methods for interpreting them. They point out the complete lack of experimental data for the unloading behaviour or materials which have undergone a shock-induced phase transformation.

Johnson (1972a) has discussed in general the complete equilibrium thermodynamic surface and also physical models of the kinetics of shock-induced phase changes. These reviews should provide an excellent background to the problem.

This section will be primarily concerned with results reported within the past several years.

6.3.1   *Iron alloys*

There is still much interest in the shock response of iron and various alloys. The high-pressure phase diagram for iron was established by both dynamic and static experiments. The interesting history of this effort by many separate investigators was summarized by Bundy (1965). The low-pressure, room-temperature phase is body-centred cubic (α), the high-pressure, moderate-temperature phase is hexagonal close-packed (ϵ) and the high-pressure, high-temperature phase is face-centred cubic (γ). The triple point was found to be at 110 kbars and 500°C. However, the recent pressure scale recalibration by Drickamer (1970) and recent work by Wong *et al.* (1968) suggest that this phase diagram may be in error. Wong *et al.* (1968) also presented evidence that the transition may be much more complicated. The end result of the 130 kbar transition may be a mixture of γ and ϵ phases with an intermediate metastable body-centred tetragonal phase.

Andrews (1973) has developed an empirical equation of state for the alpha and epsilon phases of iron. This equation is thermodynamically consistent and was fitted to all available experimental data.

Gust and Royce (1970a) report that transition pressures for Fe-Cr alloys vary continuously from the known 130 kbar value for pure iron to 230 kbar for 30 per cent Cr content. Alloys of the Fe-Cr-8Ni series exhibit almost linear decrease in transition pressure from 120 to 80 kbar as Cr is increased from zero to 18 per cent. Loree et al. (1966a, 1966b) reported results of investigations of the shock-induced polymorphism of various iron alloys, such as Fe-V, Fe-Mo, Fe-Co, Fe-C, Fe-Mn and Fe-Ni. Nickel or manganese lower the trasition pressure; vanadium and cobalt raise it dramatically.

Loree et al. (1965) discuss the overdriving effect in dynamic polymorphism. They find that the transition pressure in alloys and steel is dependent on specimen thickness and input pressure. Warnes (1967) reports similar results for antimony, where the transition pressure varied from 180 kbar for the thinnest samples to 91 kbar for the thickest. It is likely that this time dependence could be seen in many other transitions if sufficient experiments were performed.

Rohde (1970) reports experiments to measure temperature dependence of the shock-induced reversal of martensite to austenite in an iron-nickel-carbon alloy. The transition pressure ranges from 67 kbar at 45°C to 6 kbar at 390°C.

6.3.2 *Potassium halides*

The potassium halides KCl and KBr have been studied by Dremin et al. (1965) and Al'tshuler et al. (1967) using electromagnetic particle-velocity gauges. For KCl it is found that a phase transition begins at 19 to 20 kbar. Al'tshuler et al. consider that the final states reached between 20 and 51 kbar are mixtures of the two phases. The same general pattern is seen in the KBr curve, but it is less pronounced. Dremin et al. consider that the transition begins at 24 kbar. Al'tshuler et al. recorded and analysed the rarefaction waves. For KCl they find that as the material relaxes, it remains in the high-pressure phase down to about 10 kbar. At 10 kbar nearly total transformation to the low-pressure phase takes place behind a rarefaction shock wave.

6.3.3 Silicates

Ahrens *et al.* (1969) consider various shock-induced high-pressure phases among silicate and oxide rocks and minerals of geophysical interest. Of twenty-four materials analysed, all but MgO, Al_2O_3, and MnO_2 are inferred to undergo at least one shock-induced phase transition below 800 kbar. The data, as analysed, place more realistic limits on the possible constituents of the earth's mantle.

6.3.4 Boron nitride

Atmospheric pressure, boron nitride has a structure similar to graphite. Similarly, at high static or dynamic pressures it forms diamond-like modifications of the sphalerite (cubic) or wurtzite (hexagonal) type. In static compression the transition occurs at ~ 120 kbar (Bundy and Wentorf, 1963). Adadurov *et al.* (1967) and Al'tshuler *et al.* (1967) report the beginning of the transition in dynamic compression at ~ 120 kbar. Adadurov *et al.* recovered the new phase and determined that the wurtzite structure was the only high-pressure phase present. Coleburn and Forbes (1968), on the other hand, recovered the sphalerite phase with only traces of wurtzite. The recovery experiments of Dulin *et al.* (1969) yielded both high-pressure forms, the large fraction being the wurtzite phases. Likewise, DeCarli (1967a) found both high-pressure forms in his shock experiments. In the experiments of Al'tshuler *et al.*, the BN Hugoniot was traced up to pressures of 3.5 Mbar. Between 180 and 900 kbar, the results indicate the existence of relaxation phenomena due to the finite phase-transition time.

6.3.5 Carbon

The transformation of graphite to diamond in shock waves has been well described in the literature (DeCarli and Jamieson, 1961; Hamann, 1966; DeCarli, 1967b). Alder and Christian (1961) reported a second transition in graphite beginning at 600 kbar. They suggested that the discontinuity in the P-V Hugoniot represented a transition from diamond to metallic carbon. However, Pavlovskii and Drakin (1966) studied the shock compression of graphite up to pressures of 1 Mbar and found no discontinuity in the Hugoniot above the graphite-diamond transition. They suggest that Alder and Christian's data are in error, probably because their sample thicknesses were too great, causing the parameters of the shock wave in graphite to be distorted by relaxation waves. Further work by Dremin and Pershin (1968) and by McQueen and Marsh (1968) describes the dynamic graphite-to-diamond transition more accurately. These authors also found no evidence of a further transition.

6.3.6 *Silicon and germanium*

The first quantitative results on the dynamic compression curve of silicon were reported by Al'tshuler (1965). He reported the HEL to be about 35 kbar and a phase transition at about 100 kbar. Later, Pavlovskii (1968) reported a series of experiments shocking silicon along the [111] axis. He found the HEL to be about 40 kbar and a phase transition of 112 kbar. He reports indirect evidence for a possible further transition to a metallic phase above about 200 kbar.

Gust and Royce (1970b, 1971a) conducted further work on single crystal silicon and germanium using inclined mirrors and prisms. Four-wave shock structures were reported for [110] and [111] Si. Their results are given below in Table 3.

TABLE 3

Axial yield strengths and phase-transition stresses

Material	HEL kbar	Volume cm^3 g^{-1}	First transition stress kbar	Volume cm^3 g^{-1}	Second transition stress kbar	Volume cm^3 g^{-1}
[100]Si[a]	92 ± 10	0.406	none observed		140 ± 4	0.385
[110]Si	50 ± 5	0.419	103 ± 7	0.398	128 ± 5	0.385
[111]Si	54 ± 3	0.418	101 ± 3	0.400	137 ± 5	0.383
[100]Ge[b]	58 ± 8	0.179	117 ± 10	0.1663		
[110]Ge	48 ± 14	0.182	119 ± 13	0.1660		
[111]Ge	47 ± 10	0.182	125 ± 15	0.1657		

[a] Initial volume 0.429 cm^3 g^{-1}.
[b] Initial volume 0.1875 cm^3 g^{-1}.

Germanium exhibits an HEL and a phase transition when shock loaded along the [100], [110], and [111] direction (Gust and Royce, 1972). The results are shown in Table 3.

6.3.7 *Rare earth and alkali-earth metals*

The physical and chemical properties of the rare earth metals are quite similar, as the outer electron shells remain unchanged with increasing atomic number. In particular, their space lattices consist of variants of closest packing. The appearance of denser phases upon compression, then, is evidence of a change in the electron distributions.

Al'tshuler *et al.* (1966, 1968) investigated the compressibility of five lanthanides, La, Ce, Sm, Dy, and Er, up to 3.5 Mbar pressure. They report kinks in the Hugoniots signifying transitions in La at ~ 222 kbar and in Sm, Dy, and Er in the range from 590 to 740 kbar. They suggest that a transition in Ce may occur at a pressure too low to have been detected in this investigation.

Bakanova *et al.* (1970) have studied hafnium, europium, and ytterbium in a wide range of pressures from 100 kbar to 3–4 Mbar. They find that at certain critical pressures, the Hugoniots of these metals exhibit kinks which are accompanied by a discontinuous decrease in compressibility. These kinks are attributed to electronic transitions, i.e. shifts of s-electrons to inner d-levels.

Nine rare earths, plus Yb and Sc, were studied by Duff *et al.* (1968). Their data and those of the Soviet workers are in substantial agreement. Duff *et al.* suggest that at relatively low pressures the 5d-electron in the rare earths is promoted to the unfilled 4f-level. The rare earths would then become s-bonded and more compressible. The decrease in compressibility at high pressures is thought to be caused by repulsion of the closed 4p-xenon shells. Royce (1967) and Keeler and Royce (1971, pp. 95–106) discussed these processes in detail. The Soviet authors propose that the kinks in the Hugoniot result from promotion of an electron to a d-level in a second-order phase transition.

Bakanova and Dudoladov (1967) have detected electronic transitions in calcium and strontium which are presumably the result of a transition of s-electrons to unfilled d-levels.

The electronic structure and compressibility of metals at high pressures has been reviewed by Al'tshuler and Bakanova (1969). They discuss the influence of electrons on the Hugoniot at extreme pressures. Knowledge of the Hugoniot in this region is important in order to extrapolate with some accuracy into the region where the Thomas–Fermi model is applicable. Trunin *et al.* (1972) have measured the densities of iron, copper, and cadmium at extremely high pressures and have shown that the Thomas–Fermi model is applicable for these metals above 150–200 Mbar.

3.6.8 *Water*

In static compression, water is transformed into various crystal modifications of ice (Bridgman, 1942). In dynamic experiments, Walsh and Rice (1957) saw no sign of a transition in water. Al'tshuler *et al.* (1958) reported a transition in water at 115 kbar, but further experiments by Zel'dovich *et al.* (1961) showed no transition or change in transparency of water to 300 kbar. Recently, Kormer *et al.* (1968) showed that the water–ice VII phase transition can occur in a double shock experiment if the first shock is in the range 20 to 40 kbar. The transition is reported to be complete within 10^{-6} to 10^{-7} s and can be observed by the loss of transparency of the compressed layer and by the appearance of strongly diffused scattering of light. The transition does not occur in water compressed only once. The authors explain their results by the lower shock temperature reached in a double shock experiment and by time relaxation in the production of the ice.

3.6.9 *Teflon*

Champion (1971) has shocked Teflon (polytetrafluoroethylene) to stresses between 2.5 and 25 kbar. His measurements show a phase transition occurs at 5.0 ± 0.2 kbar with an associated volume change of about 2.2 per cent. The transmitted stress profiles show peak stress attenuation and suggest a viscoelastic behaviour for Teflon.

3.6.10 *Melting during shock compression*

Melting during shock-wave compression has been considered theoretically by Urlin (1966) and Horie (1967). Two essential questions are important with regard to melting in the shock wave: (1) Is melting possible in a shock compression with a duration of the order of microseconds? (2) How can information about melting be extracted from shock-wave data? Polymorphic transitions are known to occur during shock compression and Johnson and Mitchell (1972) have shown that crystal reordering sufficient to form an X-ray pattern can take place in microsecond times so that there would seem to be enough time for the melting process. Melting during shock loading is difficult to detect since it does not produce a double shock structure as is observed for polymorphism. Depending on the slope of the melting curve, there may be a change in the slope of the shock velocity versus particle-velocity data. However the most likely observable manifestation of melting would be in the temperature versus stress data. Unfortunately at present, it is possible to measure temperature only in transparent materials during shock-wave passage (see section 5.5).

Kormer *et al.* (1965) were able to determine the melting curves of NaCl and KCl during shock-wave loading by measuring the shock temperatures. They found that NaCl commenced melting at about 540 kbar and was completely melted by 700 kbar, and that KCl began to melt at about 330 kbar and was completely melted by 480 kbar.

Belyakov *et al.* (1967), using flash X-ray techniques, have observed the melting of lead during shock loading. They found that a stress of about 230 kbar was sufficient to cause lead to melt during shock compression.

Ahrens (1972) investigated the effect of initial porosity on the post shock melting and vaporization of several metals (Ba, Sr, Li, Fe, Al, U, and Th). He found that for the less compressible of these metals an optimum initial specific volume exists such that the total entropy production, and therefore the amount of liquid metal or metal vapour, is a maximum. This optimum initial volume is about 1.4 to 2.0 times crystal volume.

Naumann (1971) has developed an equation of state for metals which can be extended in a thermodynamically consistent manner to the vapour phase.

6.4 MAGNETIC PROPERTIES

Shock compression strongly affects the magnetic properties of materials. A well-known effect is the transition from ferromagnetic to paramagnetic behaviour, which can accompany either a first-order or a second-order phase transition. Curran (1961) observed a pressure-induced second-order magnetic transition in Invar (Fe-36Ni), and Graham *et al.* (1967) describe a similar magnetic transition in Fe-30Ni. A similar explanation is given by Clator and Rose (1967) for a demagnetization signal in Fe-36Ni shocked to 250 kbar.

Royce (1966, 1968) reports that shock compression of nickel ferrite causes nearly complete demagnetization for peak shock stresses between 43 and 430 kbar, whereas far less demagnetization was expected except for the highest stresses reached. He proposes that stress-induced magnetic anisotropy rotates the direction of magnetization. Seay et al. (1967), Gourley and Gourley (1968), and Shaner and Royce (1968) report evidence in ceramic manganese-zinc ferrite, Armco iron, and ytterbium-iron garnet, respectively, supporting Royce's proposal.

Graham et al. (1967) made a quantitative measurement of the ferromagnetic Curie temperature of 25°C at 22.6 kbar in Fe-30Ni alloy. They report agreement with values predicted from hydrostatic measurements made below 5 kbar on similar alloys. Graham (1968) made a relative measurement of change in magnetization for Invar (Fe-34.8Ni) above 30 kbar and reports agreement with lower pressure hydrostatic data on similar alloys. Further study by Wayne (1969) was made of saturation magnetization of Ni-Fe alloy as function of stress, both hydrostatic and shock.

Wong (1969) studied the α to (ϵ, γ) transition in iron by observing eddy current spikes in double-shock experiments. The disappearance of the spike between 150 and 180 kbar indicates the iron is no longer magnetic and the transition is complete. Royce (1968) also performed experiments on shock-induced demagnetization of iron, using a different technique. Royce measured the shock-induced reduction in magnetic flux by pickup coils wound on one arm of a magnetic circuit, which includes the sample between two ferrite arms and a permanent magnet. His results showed complete demagnetization of iron at $\sim$ 180 kbar and somewhat less demagnetization as the final pressure was lowered towards the 130 kbar phase transition, indicating that the transition might not be complete in this region. Royce's data and that of Wong (1969) are in general agreement.

Mitchell and Keeler (1967) and Keeler and Mitchell (1969) report demagnetization in iron at pressures above 130 kbar observed as a sharp negative pulse associated with a rapid decrease in magnetic susceptibility, whose decay is governed by conductivity of the iron. The authors report this as strong evidence of the nonmagnetic nature of ϵ-iron. Partial demagnetization below 130 kbar and as low as 50 kbar suggests either partial phase transition from α to ϵ iron or a change in magnetic properties not in agreement with the theoretically predicted values. More detailed work is required to explain the magnetic behaviour of iron below 130 kbar.

Grady (1972a) has devised a simple method for studying magnetic materials during shock-wave loading by projectile impact. Using this method he has studied the shock-induced anisotropy in yttrium iron garnet (Grady et al., 1972). The analysis of the experimental data is based on a theory for single crystals which is then applied to individual crystallites in polycrystalline YIG (Grady 1972b). From his work on YIG, Grady concludes that the independent grain theory is more representative of the data than previous assumptions of interacting grain behaviour. He also finds a negligible contribution from exchange and demagnetizing effects.

6.5 DYNAMIC YIELDING

In section 3.5 dislocation models were discussed to show how macroscopic plastic deformation could be related to microscopic material properties. In this section, we will discuss the results of experiments in which dislocation motion has been applied to dynamic yielding. Specifically, we will discuss stress relaxation, precursor decay, work hardening and the Bauschinger effect, recent work on single crystals, and rate effects that occur during stress release.

When a material is loaded dynamically by plane-wave impact or by high explosive, very high strain rates are imposed and a state of uniaxial strain is produced in the material. Because the strain is uniaxial, the material undergoes a volume change in the direction of loading and in addition a change in shape (i.e. a shear). The shear stresses present are inclined to the shock front, and their magnitude is limited by the shear strength of the material. When the magnitude of these shear stresses exceeds a value characteristic of the particular material, resistance to shear deformation cannot be maintained and the material yields. This yielding may be either plastic or brittle. The magnitude of the normal stress at which yielding occurs in a dynamic uniaxial experiment is called the Hugoniot elastic limit (HEL). For stresses not too far in excess of the HEL, the stress wave will separate into two parts, an elastic precursor whose amplitude is equal to the HEL and whose velocity is close to the longitudinal elastic-wave velocity, and a slower deformational wave whose velocity depends on the stress.

In some materials, the HEL agrees with the value predicted from the yield stress obtained in a quasi-static experiment (Jones et al., 1962). For other materials, the HEL can be very much greater than the value predicted from quasi-static data. This difference has been attributed to strain-rate effects (Dieter, 1961; Jones *et al.*, 1962). There are two other strain-rate effects associated with dynamic yielding: (1) the relaxation of stress immediately behind the elastic wave front; (2) a decrease in the amplitude of the HEL with increasing propagation distance. An excellent summary of HEL measurements was given by Jones and Graham (1971) and McQueen *et al.* (1970). Measurements of HELs for ceramics and high-strength materials were given by Ahrens *et al.* (1968), Lyle (1968), and Gust and Royce (1971a, 1971b), and Gust *et al.* (1973). HEL measurements as a function of temperature for tantalum have recently been reported by Rohde and Towne (1971). Gilman (1968) published an excellent review of the work done in relating dislocation dynamics to dynamic yielding in polycrystalline materials. J. N. Johnson *et al.* (1970) reported on work done on single crystals.

6.5.1 *Stress relaxation*

Stress relaxation behind the elastic wavefront has been observed in iron and iron alloys (Jones *et al.*, 1962; Taylor and Rice, 1963; Jones and Holland, 1964; Holland, 1967; Butcher and Munson, 1968; Rohde, 1969b) and in aluminium (Barker *et al.*, 1966; Butcher and Munson, 1968). Other materials in which stress relaxation occurs are single crystals of beryllium (Taylor, 1968), cadmium sulphide (Kennedy and Benedick, 1966), sodium chloride (Murri and Anderson, 1970), copper (Jones and Mote, 1969), germanium (McQueen, 1964; Graham and Ingram, 1968), and quartz (Wackerle, 1962; Fowles, 1967).

Stress relaxation is attributed to a sudden increase in the mobile dislocation density when yielding occurs. Initially all the strain is elastic, but as yielding occurs and the wave moves into the material the strain is relaxed by dislocation motion. The stress then decays towards the yield point. If stress relaxation is to be observed, the stress in the material behind the wave must relax faster than the stress at the wavefront otherwise the entire elastic wave would decay at the same rate. For stress relaxation to occur as we have described, the mobile dislocation density must increase dramatically behind the wavefront. The increase in mobile dislocation density may come about by unlocking previously pined dislocations or by creation of new dislocations during plastic straining, or both. This model of stress relaxation was suggested by Barker *et al.* (1966). Computer results (Butcher and Munson, 1968) based on this model are in accord with some experimental observations. Holland (1967) gave further support to this model. He showed that stress relaxation can be suppressed by increasing the mobile dislocation density before dynamic loading. Apparently, if the initial dislocation density that can be mobilized by dynamic loading is great enough to accommodate the imposed strain without requiring additional mobile dislocations, then stress relaxation will not occur. Another result of increasing the initial dislocation density is to reduce the yield stress. Jones and Mote (1969) prestrained copper single cryssals before loading and observed essentially zero yield stress and ramp-like elastic waves. They found similar behaviour in polycrystalline samples.

6.5.2   *Elastic precursor decay*

The decay of the amplitude of the HEL with propagation distance has been observed in iron and iron alloys by several workers (Jones *et al.*, 1962; Taylor and Rice, 1963; Ivanov *et al.*, 1963; McQueen, 1964; Peyre *et al.*, 1965; Butcher and Munson, 1968). It has also been observed in aluminium (Novikov *et al.*, 1966; Anderson *et al.*, 1967; Karnes, 1968), in tantalum (Gillis *et al.*, 1971), in quartz (Wackerle, 1962), in lithium fluoride (Asay *et al.*, 1972a; see below under single crystal), in antimony (Warnes, 1967), and in brass (Novikov *et al.*, 1966). The yielding may proceed by slip or by twinning. Taylor (1965) employed the theory of dislocation dynamics and the method of characteristics to study the experimental results for Armco iron. With this model, he was able to account reasonably well for the observed precursor decay in Armco iron. He assumed that the resolved shear stress was along planes oriented at 45° to the shock front. Johnson (1969) extended Taylor's model by averaging the shear stress over a random distribution of glide systems. Johnson and Band (1967) used a model based on the artificial viscosity method to study the precursor decay in iron. Their calculations agree qualitatively with experimentally measured profiles in iron and aluminium. Johnson and Barker (1969) have used steady propagating plastic wave analysis to study the rate-dependent behaviour in 6061-T6 aluminium. These steady-propagating plastic waves result from a balance between the nonlinear material behaviour which tends to cause the wavefront to steepen and the rate-dependent material properties which tend to spread out the wavefront. Similar models have also been applied to polycrystalline aluminium by Butcher and Munson (1968).

Chang and Horie (1972) have studied the problem of data analysis of shock wave propagation using dislocation dynamics. They discussed the difficulties produced by oscillations in the numerical solutions of the equations. They also note that conclusions drawn by steady-profile analysis may not be carried over to unsteady wave profiles.

Rohde (1969b) studied the effect of initial temperature on the yield behaviour of iron. He found that the yield stress was independent of initial temperature between 76 and 573 K. Gillis (1970) has discussed the temperature dependence of constitutive relations with particular reference to iron. Rhode further observed that twinning occurred at stress levels slightly above the HEL but did not occur at lower stresses. Thus the mechanism of dynamic yielding in iron is by the motion of twinning dislocations.

The effect of grain size on dynamic yielding was studied by Jones and Holland (1968). They found that grain diameters between 9.4 and 70.3×10^{-3} mm have no effect on the dynamic yielding in mild steel. They conclude that, during yielding and initiation of plastic flow, the dislocations do not move over great enough distances to encounter grain boundaries. This behaviour is in contrast to quasi-static yielding, where dislocation pileup at grain boundaries is important (see Russell et al., 1961).

Asay et al. (1972b) have developed a theoretical procedure for reducing the observed precursor decay data to obtain direct information about the material constitutive relations.

6.5.3 Work hardening and Bauschinger effect

Two other phenomena associated with dynamic yielding that have been observed are work hardening and the Bauschinger effect. When yielding occurs, the magnitude of the shear stress decreases, and for elastic–perfectly plastic behaviour, a constant value independent of stress is obtained. However, if, after the decrease at yielding, the shear stress above the HEL increases with increasing compression, then work hardening is occurring. This work hardening is the result of dislocation entanglement and pinning so that an increasingly greater stress is necessary if the plastic flow is to continue. Although work hardening might be expected to be quite important, experimental results to date indicate it is a small effect. Perhaps the reason is that, as shown by Jones and Holland (1968), the dislocations do not move very far during the yielding and subsequent plastic flow, so that dislocation interaction is minimized. Weidner and Royce (1969) suggested that the discontinuity at 29 kbar in [100] NaCl observed by Larson (1965) is the result of work hardening of the primary slip system so that a secondary system is activated.

Ma (1972) has suggested that materials may possess a "super shear stress" during shock loading. He suggests that when the shock-wave velocity in a solid reaches sonic velocity, the effectiveness of stress relaxation by plastic flow may become negligible. Therefore when the cohesive shear strength is reached during shock loading a kind of "elastic slip" without active participation of dislocation motion occurs.

The Bauschinger effect has been observed in several shock-loaded metals. Jones and Holland (1964) reported measurements in mild steel. In their experiments, some specimens were given 20 per cent tensile cold work before compressive loading. The HEL for the cold-worked specimens was about 8.8 kbar compared with 15.7 kbar for the annealed samples. The necessity of including the Bauschinger effect to account for experimental observations made during stress release has been noted by several workers. This problem is discussed below in connection with rate effects occurring during stress release.

Orowan (1959) explained the Bauschinger effect with the following model. A dislocation line will move along a slip plane under the action of a shear stress until it is blocked by obstacles strong enough to resist shearing. When the loading stress is removed, the dislocation line will not move appreciably unless very strong back stresses are present. However, when the direction of loading is reversed, the dislocation line can move a substantial distance at a low shear stress because the obstructions to the rear of the dislocations are not likely to be as strong or dense as those directly in front. As the dislocation line continues to move, the impedance to its motion increases and the shear stress increases with strain. The above result of Jones and Holland are in agreement with this model.

Dispersive elastic waves were observed in polycrystalline beryllium by Taylor (1968) and Christman and Froula (1970). The magnitude of the HEL also varies greatly with crystallographic direction. This pronounced anisotropy of the HEL may cause the observed dispersion and smearing of the elastic wavefront in the polycrystalline material.

6.5.4   *Recent results for single crystals*

The use of single crystals allows the investigator to determine accurately the resolved shear stress on specific slip systems and to load samples in specific crystal directions. Results of HEL measurements have been reported in a number of single crystals (see McQueen *et al.*, 1970; Murri and Anderson, 1970; J. N. Johnson *et al.*, 1970; Jones and Graham, 1971; Asay and Gupta, 1972; Asay *et al.*, 1972a, 1972b).

A common result of the work on single crystals is the dependence of the HEL magnitude on the crystal direction of dynamic loading. The variation in HEL magnitude in many cases is quite large; for example, in beryllium, the HEL parallel and perpendicular to the *c*-axis differ by a factor of ten (Taylor, 1968) and in NaCl, the HEL in the [100] and [111] directions differ by a factor of about 27 (Murri and Anderson, 1970). Murri and Anderson (1970) and J. N. Johnson *et al.* (1970) were able to account for this anisotropy in the HEL magnitude by considering the resolved shear stress produced on the available slip systems by the dynamic loading. Jones and Mote (1969) calculated the precursor decay for wave propagation in the [100], [110], and [111] directions in face-centre-cubic crystals (copper in this case), and J. N. Johnson *et al.* (1970) extended the analysis to an arbitrary crystal structure. Johnson *et al.* specifically applied their theory to fcc, bcc, rocksalt and hcp crystal structures. To agree with the experimental observations, however, initial mobile dislocation densities must be one to three orders of magnitude greater than the measured total dislocation density before loading. Recently, Johnson (1970) and Johnson and Rohde (1971) published preliminary results in which twinning during yielding was considered. They found these results could adequately represent the experimental results on iron.

Asay *et al.* (1972a, 1972b) and Asay and Gupta (1972) have studied the effect of impurities and impurity clustering on the elastic precursor decay in LiF. They found that both divalent cations and interstitial ions can cause variations in the precursor decay. Further they found that the rate of attenuation of elastic waves was reduced for a given defect concentration if the defects were clustered. The clustering was shown to affect the yield behaviour differently at low and high strain rates.

Graham and Brooks (1971) have studied the effects of large anisotropic compression in sapphire. They conclude that sapphire suffers
a substantial loss of shear strength when it is shocked above the HEL.
They also suggest that different shear-failure mechanisms may be
encountered in a given crystal shock loaded along different crystallographic axes.

Most experimental work has been carried out on either isotropic
materials or on anisotropic materials shocked-loaded in a direction such
that conditions of pure longitudinal motion occur. Johnson (1971) has
theoretically studied shock propagation by planar impact in linearly
elastic anisotropic materials of arbitrary crystallographic orientation.
He predicts that multiple wave propagation and transverse particle
motion can occur. For impact in y-cut α quartz these effects should be
large enough to measure. He has extended this analysis to include
elastic–plastic behaviour and finds that transverse particle motion and
complex wave propagation behaviour can occur (Johnson, 1972b).

6.5.5 Rate effects during release

Rate effects during stress release have been studied by a number of
workers. Fowles (1960) and Morland (1959) predicted that the first
relief wave to overtake a shock wave should travel with the elastic
velocity in the compressed material. Experimental confirmation of these
predictions was obtained by Al'tshuler et al. (1960). They were able to
explain their experimental results in iron and copper by assuming two
sound speeds in the compressed materials. Weak rarefaction disturbances propagated through the compressed material with the elastic
sound speed, and disturbances with significant strength travelled with a
slower "plastic" wave speed which corresponded to the slope of the
Hugoniot curve. Curran (1963) observed premature attenuation of a
200 kbar shock in aluminium by a following rarefaction wave. He
attributed this early attenuation to an unexpectedly large elastic
release wave preceding the larger plastic release wave. Erkman and
Christensen (1967) have also observed that the initial attenuation of a
shock wave in aluminium is the result of an elastic rarefaction wave.
Erkman and Christensen found that the data for 2024–T351 aluminium
indicated a Bauschinger effect while the behaviour of annealed 1060
aluminium was more nearly elastoplastic. All of the above work was
done at stresses of several hundred kbar where it had been previously

thought that shock compressed materials would be truly hydrodynamic. Apparently the effects of material rigidity must be taken into account even at these stresses.

Barker *et al.* (1964) studied the rate effects during loading and unloading in 6061–T6 aluminium at stresses to 22 kbar. They found that the Bauschinger effect must be included in order to explain their experimental results. They also observed the propagation of an elastic compression stress wave in the aluminium after the aluminium was dynamically stressed beyond the yield point and held at constant strain for about 2 μs.

More recently, several investigators have used manganin stress gauges to study the unloading behaviour of materials. Fuller and Price (1965, 1969) have studied aluminium and magnesium at stresses up to 200 kbar. Kusubov and van Thiel (1969) have studied the release wave profiles in aluminium between 30 and 130 kbar. The results from all these studies indicate that the stress release wave has an elastic and a plastic part in contradiction to a purely hydrodynamic model. However significant dispersion is observed in the elastic release wave, indicating that a more elaborate model than the elastic-perfectly plastic model is needed to explain the experimental results. Even the inclusion of a Bauschinger effect is not enough to allow a complete fit of the stress-release profile.

The advent of in-material gauges (stress and particle velocity) and the associated Lagrangian analysis of the gauge profiles promises to be a major tool in understanding the dynamic yield behaviour in both loading and unloading.

6.6   RESIDUAL EFFECTS

The problems of containing a shock-loaded sample for subsequent examination can be difficult. The main difficulties are concerned with the rarefaction waves that release the compressive stress. The rarefaction waves must be controlled so that they do not intersect in such a way as to cause the specimen to be torn apart by tensile stresses. This is usually achieved by an arrangement of momentum traps whereby the specimen is in unbonded contact with material of similar shock impedance (Doran and Linde, 1966; Linde and Doran, 1966). When a tensile wave is propagated back towards the specimen from a free surface, it causes separation of the layers of the momentum-trapping assembly, which cannot support tensile stresses. The specimen itself unloads to zero stress without undergoing tension.

In the following sections we discuss a number of recent observations made of shock-loaded material. A major field has been that of studying, through the use of an electron microscope, the microstructure of shock-loaded metals. Results are discussed in terms of such mechanisms as twinning and production of dislocations and point defects. Studies have also been made of residual changes in magnetic, optical, and mechanical properties. In addition, deformation and changes in the crystalline lattice by shock-loading have been investigated by measuring changes an electrical resistivity, X-ray diffraction, and stored energy.

6.6.1   *Microstructure*

With electron microscopy, many of the microstructural details of shock-deformed metals and other materials can be examined. Murr and Rose (1968) studied the annealing response of stainless steel shock loaded to 120 to 1200 kbar. They found significant differences, which increased with the stress, between the shock-deformed steel and cold-rolled steel. The shocked material showed little recrystallization, with annealing characterized by prominent recovery and grain growth.

Shock-loaded stainless steel of intermediate stacking fault energy was studied by Murr and Grace (1969) and by Marsh (1970) over the range 30 to 360 kbar. In contrast to the behaviour under conventional working conditions, shock deformation produces planar arrays of dislocations at all pressures. Twinning occurred at pressures above 100 kbar. The tangled dislocation cell structure seen in final specimens was attributed to modification of planar arrays in the more diffuse rarefaction zone. Murr et al. (1971) studied the effects of explosive forming of stainless steel.

The rarefaction waves that follow shock compression can strongly affect the resulting microstructure. Mahajan (1969) observes a twin-complementary twin pair in iron shocked to $\sim$ 100 kbar. He explains that this structure requires both compressive and tensile stresses during deformation. The tensile stresses arise from interaction of rarefaction waves.

Rhode et al. (1968) examined specimens of a martensitic Fe-30 Ni alloy that had been shocked to peak stresses of 18 to 100 kbar. They concluded that shock loading below the 80 kbar phase transition reported by Loree *et al.* (1966b) initiates a martensitic reversal to the austenite phase. The amount of the reversal that occurs increases with increasing shock stress. At lower stresses, 18, 50, and 70 kbar, the reversal occurs only in limited regions resulting in a bonded microstructure. The sample shocked to 100 kbar showed a more general reversal. They conclude that it is the shear components present in a shock wave that are important in the reverse martensite transformation at low stresses. A detailed crystallographic study of the shock-induced phase transformations in iron alloys is presented by Bowden and Kelly (1967).

In general, metals and alloys with low stacking-fault energies deform readily by the introduction of mechanical twins as well as by production of dislocations and point defects. As the stacking-fault energy increases, the amount of material twinned during shock deformation decreases. Rose *et al.* (1967), Rose and Inman (1969), and Murr and Foltz (1969) studied the shock response of nickel, which has high stacking-fault energy, over the stress range of 280 to 1500 kbar. Mechanical twinning was observed at the lower pressure ranges, but when the stress was raised to 820 kbar and above, mechanical twinning could not be observed. Thermal effects dominate the high-pressure microstructure, with the microstructure of the specimen shocked to 1500 kbar characteristic of the fully annealed state.

Murr *et al.* (1970) performed shock-loading experiments to compare the substructures of nickel, TD (thorium dioxide)-nickel, Chromel-A (80Ni-20Cr), Iconel 600 (Ni-15Cr-7Fe), and TD-NiCr on the basis of similar compositions and stacking-fault energies. Well-defined dislocation cells having average diameters of 0.8, 0.3, 0.2, and 0.1 μm at stresses of 80, 180, 240. and 460 kbar, respectively, were observed in pure nickel by transmission electron microscopy. Dislocation cells were not so well developed in annealed TD-Ni because of the presence of ThO_2 particles 340 Å in diameter. The ThO_2 particles also inhibited twinning in both TD-Ni and TD-NiCr. Planar dislocation arrays were observed in chromel-A and Inconel 600 but were generally not observed in TD-NiCr shocked to 80 and 180 kbar. The authors conclude that the major part of the residual hardening in these shock-loaded metals and alloys can be attributed to production of matrix dislocations. Variations in the hardening characteristics can be attributed to the difference in stacking-fault energy between nickel and TD-Ni and the NiCr alloys.

The annealing response of shock-loaded nickel was studied by Murr and Vydyanath (1970a, 1970b) and Trueb (1969). Murr and Vydyanath suggest that vacancies form a prominent portion of the high-pressure microstructure, contribute to shock strengthening, and are the single most effective mechanism involved in activation of catastrophic recovery at elevated temperature.

Trueb studied the microstructure of polycrystalline nickel after explosive loading and subsequent heat treatments between 600 and 780°C. At the lower peak stresses (70 and 320 kbar), the deformation substructures resembled those caused by cold working, but recovery was mainly dislocation migration and recombination. Polygonization, nucleation, and grain growth were not significant. Specimens shocked to 1000 kbar showed a high density of dislocations, point defect clusters, and microtwins in complex patterns. Heat treatment causes nonuniform polygonization, nucleation, and grain-boundary migration, resulting in lower dislocation density after annealing.

Grace *et al.* (1968) and Murr and Grace (1969) examined 70/30 copper-zinc alloy (α-brass) shock-loaded to 50 to 400 kbar. Three types of defect structure were noted: dislocations, stacking faults, and twinning. The absence of dislocation cell structure is attributed to the low stacking-fault energy of this alloy. At the higher pressures reached, imperfect fault bands were observed with a predominant twin orientation.

Grace and Inman (1970) studied shock-loaded α-brasses of varying stacking-fault energies to determine the relationship between stacking-fault energy and the formation of dislocation configuration in fcc metals. The dislocation configurations observed after 55 kbar did not vary significantly from those produced by conventional means of deformation. A transition from planar dislocations at low stacking-fault energies to a dislocation cell structure at higher stacking-fault energies was observed at 25–36 erg cm^{-2}.

The microstructure of shock-loaded copper was studied by Cohen *et al.* (1966), George (1967), Brillhart *et al.* (1967), Carpenter (1968), Murr and Grace (1969), and Higgens (1971). George found the dislocation density to be very strongly dependent on shock pressure over the range 90 to 150 kbar. Brillhart *et al.* and Carpenter, on the other hand, found little change over this range. Carpenter notes that shocking at all pressures decreased the dislocation damping from that observed for unshocked copper. He suggests that, while shock pressure is an important parameter, the rate at which deformation takes place may be more important in determining the final internal dislocation structure in explosively shocked materials. Cohen *et al.* (1966) discuss twinning in shock-loaded copper. Higgens examined the substructure of copper after shock-loading to 433 kbar. He finds that conventional recrystallization is the dominant process during annealing.

Mikkola and Cohen (1966) studied the substructure of Cu_3Au after shock loading and after ordinary tensile deformation. Both types of deformation created {111} and {001} domain boundaries in ordered Cu_3Au, but the amounts were greater after shock loading. Similarly, the change in degree of long-range order is greater in shock loading. The more effective disordering in shock loading is attributed to more random dislocation distribution. Twins were identified in both ordered and disordered shock-loaded specimens.

Shock deformation of polycrystalline aluminium was studied by Rose and Berger (1968). They found that aluminium does not exhibit a dislocation cell structure after shock loading to pressures up to 150 kbar, in contrast to copper and nickel, which do under similar loading conditions. Numerous dislocation loops are formed by interstitial clustering, collapse of vacancy clusters, and dislocation reactions. Structures comparable to those produced by shock waves are not formed when aluminium is slowly deformed at room temperature to the same plastic strain. However, at low temperature and large strains, structures similar to those produced by shocking are observed after slow deformation.

Ma *et al.* (1969) studied the effects of explosive loading (60 to 70 kbar) on the structure of a magnesium single crystal. After loading, the specimen consisted of very small grains of about 500 Å. The authors suggest that these subgrains result from instantaneous polygonization of the strained and rotated lattice in the course of explosive loading.

The operative slip systems in shock-loaded MgO crystals were studied by Klein and Edington (1966). Results are explained in terms of multiplication and movement of dislocations, rather than nucleation of disclocations at the shock front.

Shock-synthesized diamond formed from the graphite in cast-iron has been studied by Trueb (1968). The microstructure is found to be very different from that of either natural diamond or diamond produced by the Allied Chemical process (P. S. DeCarli, U.S. Patent No. 3238019, March 1966). The material consists of two different forms:

compact aggregates of crystallites with a strong preferred orientation, and single crystals. The hexagonal diamond phase was found in the form of both single crystals and stacking faults within a matrix of cubic material.

In later work (Trueb, 1971) diamond was explosively synthesized from metal–graphite mixtures at high temperatures, moderately high pressures, and relatively long times at pressure. The diamond was found to be entirely polycrystalline with a microstructure quite different from other known types of shock-synthesized diamond. Only the cubic phase was found, indicating that the hexagonal phase is not stable at high temperatures.

Trueb (1970) also describes the microstructure of graphite remaining after explosive experiments designed to investigate the conditions for producing diamond. The remaining graphite in some instances had become darker and harder than the starting material. Trueb concludes that the darker graphite was transformed into diamond by the explosive shock but was subsequently regraphitized at the high temperature of the immediate postshock period.

The study of shock effects in rocks and rock-forming minerals is a specialized field beyond the scope of this review. Much of the work is oriented towards describing and interpreting the nature of the lunar surface and the results of meteorite impacts. Recent reviews can be found in Chao (1967), and in the volume edited by French and Short (1968).

6.6.2 *Magnetic effects*

The first study of residual magnetic effects in shock-deformed materials seems to have been by Rose et al. (1969). They measured the magnetic hysteresis and the normal magnetization curves of Armco iron after shock deformation. The stress range of interest was 73 to 250 kbar. They found that all structure-sensitive properties changed, and they were able to relate these changes to the peak stress associated with the shock wave. The data were discussed in terms of magnetic domain theory, and functional expressions were obtained for the change in coercive force and for the change in area of a hysteresis loop with stress. They noted a minimum in the stress dependence of the residual magnetization near the stress for the $\alpha \rightarrow \epsilon$ phase transformation (about 145 kbar in these experiments).

Christou (1970a) measured the susceptibility and saturation magnetization in shock-loaded FeMn (92.4/7.6 wt%) between 90 and 300 kbar. Three distinct magnetization regions were found: (1) an initial increase in saturation magnetization that is less than that in the fully annealed state; (2) a rapid increase in magnetization known as the Barkhausen effect; (3) a region in which the approach to saturation occurs at higher fields than in the annealed state. The magnetization changes in these three regions were explained in terms of the alloy's dislocation and domain substructure. A density change of 3.79 per cent was measured for specimens shocked at stresses of 150 and 300 kbar, suggesting a crystallographic phase change. An observed increase in susceptibility was attributed to a shock-induced Curie transition.

Christou (1970b) measured the transverse magnetoresistivity of annealed and shock deformed Fe, Fe-7.37 wt% Mn, and Fe-30 wt% Ni. The experiments were performed in the stress range of 90 to 500 kbar. Christou plotted the magnetoresistivity results for annealed and shock-loaded iron and iron alloys on a Kohler diagram and observed that in general the deformed metals were shifted from the annealed ones. The shift in Fe could be explained in terms of the anisotropic scattering of conduction electrons by dislocations. The shift in the Fe-Mn and Fe-Ni alloys was the result of a shock-induced second-order phase transformation occurring above 90 kbar.

6.6.3 *Lattice defects and optical properties*

Kressel and Brown (1967) studied the production of lattice defects produced during shock deformation in high-purity nickel by measuring the electrical resistivity on recovered specimens. They found that the dislocation density and the point-defect concentration following shock deformation was considerably greater than they were for equivalent values of plastic strain achieved by cold rolling. Approximately eight times more vacancies are formed than interstitials, but the ratio of these two types of defects appears to be strain rate independent.

For small strains, the vacancy concentration increases linearly with plastic strain until it reaches a saturation value at large strains. This result is thought to occur because vacancies are absorbed by mobile dislocations moving in their vicinity during the deformation, and the rate of vacancy formation tends to equal the absorption rate for large strains. The major formation process for point defects is the nonconservative motion of jogs on screw dislocations and not the recombination of dislocation segments containing edge-type components.

The change in electrical resistivity was used to study the ordering kinetics of Fe-Mn alloys (Christou, 1972) and of Ni-Cr alloys (Christou and Brown, 1973).

Batsanov and Lapshin (1967) made a study of the consequences of explosive compression of pure zinc sulphide. They found a variety of effects including changes in paramagnetism, blue and green luminescence centres in different grains, and a reduction of the photoluminescence intensity of shock-loaded samples after heating. Similar work was performed on ZnS(Cu) phosphor by Nomura and Tobisawa (1965). Horiguchi (1966) found that shocking $Zn_2SiO_4(Mn)$ phosphor increased the luminescence intensity significantly.

6.6.4   *Mechanical properties*

We mention here a few recent papers that are concerned with the effects of shock waves on mechanical properties of materials. For a more extensive review, the reader is referred to Doran and Linde (1966).

Hardness is the mechanical property most commonly studied in shock-loaded materials. It has been shown to be closely related to microstructure, such as dislocation density, cell formation, twinning, and point defects (Murr and Grace, 1969). Murr and Rose (1968) find that, in stainless steel shock-loaded to pressures ranging from 120 to 1200 kbar, the increased hardness is due mainly to high dislocation densities in the pressure range 120 to 235 kbar, to twin-fault structures in the pressure range 235 to 525 kbar, and possibly to a gradually increasing point defect density above 525 kbar. The transient shock temperature rise and effective quench appear to be controlling factors influencing residual hardness at very high pressures.

Rose *et al.* (1967) studied shock hardening in nickel with peak stresses of 280, 380, and 710 kbar. They conclude that, for a given surface, the saturation value of mechanical hardness for shock-loaded nickel is essentially independent of prior cold work and that stacking faults and twin faults do not contribute significantly to the shock-hardening process in nickel. At higher pressures (Rose and Inman, 1969) thermal effects begin to become important. Between 400 and 1200 kbar, mechanical hardness slowly decreases. Beyond 1200 kbar, shock-heating causes the material to recover completely. For a shock pressure of 1500 kbar, the hardness and the microstructure are characteristic of the fully annealed state.

Murr and Foltz (1969) examined shock-loaded samples of nickel and Inconel 600 alloy (76Ni-16Cr-7Fe) containing fine precipitates. The effect of the precipitates was to enhance shock hardening by acting as sources for additional dislocations, dislocation dipoles and loops, as well as barriers to dislocation motion. The residual hardness of nickel was observed to become saturated above about 300 kbar, but that of Inconel continued to rise steadily. The differences in hardness behaviour were related to observed differences in the substructures. Transmission electron microscopy revealed a steadily decreasing cell size with increasing shock pressure in nickel, with almost no evidence of deformation twins at 370 kbar. The Inconel substructure was characterized by planar dislocation arrays including an increasing concentration of dipoles and elongated loops to 200 kbar. At 370 kbar deformation twins occupying 19 per cent of the volume were observed.

Orava (1973) has shown that the combination of explosive shock loading and ageing is an effective method of improving the tensile properties of a nickel-base superalloy.

Grace et al. (1968) investigated shock-induced hardening in α-brass. Hardness increased rapidly with applied pressure to about 200 kbar then increased at a slower rate until a plateau was reached in a pressure range of 400 to 500 kbar. The increased hardness is attributed to shock-induced dislocation and twin-fault densities.

Zukas (1966) discusses strengthening of a number of metals by shock loading. In most materials, the changes brought about by shock loading alter their microstructures and increase their hardness and strengths. Steels, which have been hardened by heat treatment, were not hardened additionally by shock loading, suggesting that similar dislocation mechanisms are responsible.

From studies of aluminium at different strain rates (3.8×10^{-3} s^{-1} and 2.5×10^3 s^{-1}), Taborsky (1969) infers that the improved mechanical properties of aluminium on shock loading are caused by changes in cell structure and by increased dislocation velocity.

6.6.5   *X-ray diffraction*

The study of X-ray line broadening in shocked specimens can yield a description of the defect structure induced by deformation. The method of analysis is discussed by Warren and Averbach (1950) and by Warren (1959). Brillhart *et al.* (1967) studied shock-loaded copper using a number of techniques, including X-ray diffraction. The authors conclude that many types of measurements are required to obtain detailed information on the structural changes. X-ray measurements are suitable for calculation of dislocation density only if electron microscopy is used to ascertain the presence, density, and arrangement of faults or twins.

Klein and Rudman (1966) used X-ray line-broadening methods and transmission electron microscopy to determine the shocked structure of MgO single crystals. The line broadening is caused primarily by small particles, of the order of 200 Å, that have their origin in regions between dislocations or small dislocation clusters. The authors conclude that the shocked structure has a dislocation density near 10^{12} cm^{-2} and that the dislocations are arranged in configurations of low lattice strain.

The explosive shock treatment of nonmetallic powders can be used to increase their sintering rate. The enhanced sintering behaviour may arise from reduction of the powder particle size or from an increase in the number of defects in the atomic lattice, or from both. Bergmann and Barrington (1966) studied the X-ray line broadening of shocked SiC and B_4C powders to determine lattice strain and crystallite size reduction. Heckel and Youngblood (1968) studied shocked MgO and α-Al_2O_3 powders.

6.6.6 *Stored energy*

Brillhart *et al.* (1967) report that stored energy in shock-loaded copper is larger than that caused by conventional deformation. A significant factor is the high density of twins. Dislocation densities from stored energy calculations are high, probably because the deformation twin boundaries present are largely incoherent.

Mikkola and Cohen (1966) conclude that the stored energy of deformation in Cu_3Au is primarily associated with the energies of the {111} antiphase domain boundaries in ordered Cu_3Au and that disordering proceeds by the cutting up of the material into smaller domains.

Grace (1969) finds that the energy stored in metal during shock loading can be determined adequately from plastic work calculations and from calorimetric data. Results suggest the ultimate strengthening of metals by shock waves is limited by the saturation of energy stored in the lattice.

6.7 SHOCK-INDUCED FRACTURE

In the case of quasi-static loading of brittle materials it is often possible to apply a simple fracture criterion, namely, that a critical tensile stress exists beyond which failure occurs. Sample geometry and the effects of elastic material properties and initial crack size can be included by determining the "fracture toughness", which is related to the maximum stress concentration at a crack tip allowable before an initially present crack propagates.

These simple criteria, however, are not compatible with the results from dynamic experiments, since it is found that there is a time dependence for dynamic failure (Smith, 1963; Butcher, 1968; Warnicka, 1968; Charest et al., 1969), i.e. the fracture criteria depend on the entire stress history. For square stress pulses this time dependence of dynamic fracture has been shown (Tuler and Butcher, 1968) to be consistent with a relationship between tensile stress σ and the time duration of tensile stress Δt. This relationship is

$$(\sigma - \sigma_0)^\lambda \Delta t = K$$

where σ_0, λ, and K are constants specific to a particular material. This equation, with $\lambda = 1$, is a simple impulse criterion. With $\lambda = 2$, it can be shown to be equivalent to an energy criterion (Jajosky and Ferman, 1969). The equation with $\lambda = 1$ can also be related to other time-dependent fracture criteria, such as stress gradient (Breed et al., 1967) and stress loading rate (Skidmore, 1967), which are consistent with data resulting from explosive tests. However, these criteria are based primarily on a simplified classification of damage into categories of no damage, damage, and failure. A model that relates fracture behaviour to detailed observations of the damage resulting from a given stress history was presented by Barbee et al. (1970, 1972), Seaman et al. (1971), and Shockey et al. (1973).

In this model the microscopic damage is treated as a new phase of the material, where the nucleation and growth of this new phase is correlated with *macroscopic* stress history and *microscopic* material properties. A necessary condition is that the number of microcracks be large enough to make a statistical approach valid. The steps of the method are: (1) shock and recover specimens; (2) quantitatively describe the size and spatial distributions of cracks; (3) achieve experimental control so that the cracks can be stopped in different stages of growth; and (4) specify the macroscopic stress and time at stress experienced at any location in the specimen. If these requirements are met and if the stress history is sufficiently uncomplicated, then it is possible to correlate the observed damage with the stress history of the specimen in such a way that nucleation and growth laws of general validity may be derived. Once these factors are known, quantitative predictions can be made of damage caused by an imposed stress history.

The stress histories are determined by a combination of experimental measurements with manganin gauges (Keough, 1968) and numerical calculations with computer codes designed for the plate-slap geometry (Seaman, 1970c).

The quantitative characterization of damage in step (2) of the above method is accomplished by dividing the sectioned and polished sample into zones corresponding to varying tensile stress durations. The cracks in each zone are then counted, and their apparent sizes and orientations are recorded. Variations of standard statistical procedures are used to map the observed surface size and orientation distributions into the actual volume distributions (Kaechele and Tetelman, 1969). This is done for several shots at varying stress levels. Thus, one obtains the number density of microcracks as a function of stress, time-at-stress, orientation, and size. (In plate-slap experiments, the stress pulses are approximately square waves, so the stress history can be completely

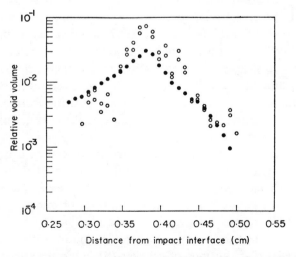

Fig. 46. Comparison of computed and experimental void volume distribution for a 9.2-kbar tension shot in 1145 aluminium. Computation (●); experimental points (○). (After Barbee *et al.*, 1970.)

specified by stress amplitude and time-at-stress.) From the number density function $N(\sigma, t, \Phi, R)$, where σ is the tensile stress, Φ is the crack orientation, t is the time-at-stress, and R is the crack radius, the nucleation rates $(\partial N/\partial t)_{\sigma, \Phi, R}$ and the growth rates $(\partial R/\partial t)_{N, \sigma, \Phi}$ can be obtained, and from these parameters damage can be predicted for given stress histories.

This approach has been successful in predicting quantitative levels of shock damage induced in samples loaded by sudden deposition of radiation from X-ray and electron beam sources, as well as by mechanical impact (Shockey *et al.*, 1973). The nucleation and growth para-

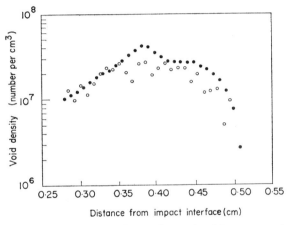

Fig. 47. Comparison of computed and experimental void concentrating distribution for the same 9.2-kbar shot shown in Fig. 46. Computation (●); experimental points (○). (After Barbee *et al.*, 1970.)

meters necessary for prediction of shock damage have been determined by the above authors for 1145 aluminium, OFHC copper, Armco iron, several grades of beryllium, and polycarbonate. Figures 46 and 47 show compared theoretical and experimental void concentrations and volume in 1145 aluminium specimens that were impacted by flying plates.

7 Comparison of static and dynamic results

One of the original motivations for shock-wave experiments, using explosives, was to obtain pressure–volume data in high-pressure regions that could not be reached with static pressure experiments. In later years, considerable overlap in experimental pressure regions has been possible with the advent of the light-gas gun for lower pressure dynamic work and with improved technology for static work enabling pressures up to about 500 kbar to be reached. Where detailed comparisons between static and dynamic measurements can be made, it is usually found that agreement is good. There are many important differences between the two kinds of experiments, however, that must be considered in interpreting the experimental results.

The most obvious difference is the length of time that a specimen can be subjected to high pressure. Shock-wave experiments are conducted in microsecond time scales. Surprisingly, this limitation does not seem to be so severe as might be imagined. Phase transitions such as in iron that are considered "sluggish" in static experiments are known to occur in shock experiments.

Another difference is in the stress distribution. In static work the stress is more nearly hydrostatic. However, at pressures above about 30 kbar the pressure transmitting media solidify, causing a poorly defined shear stress to be present. Shock measurements, on the other hand, are generally conducted in a system of well-defined one-dimensional strain. The shear stresses present in shock experiments are thought to constitute an additional driving force for phase transitions, compensating for the short time duration. Thus, a transition that is slow in static compression may still occur within a few microseconds in shock experiments.

Temperatures are more easily known and controlled in static experiments, but much higher temperatures can be reached in shock compression. In comparing static and dynamic results, the shock compression curve is usually corrected for shock heating by assuming a Mie–Gruneisen relationship (Walsh and Christian, 1955; Knopoff and Shapiro, 1969; Shapiro and Knopoff, 1969; Kompaneets et al., 1972; Davis, 1972; Grover and Rogers, 1973).

Because of the different stress distribution, materials with a high HEL may require a significant "strength-of-materials" correction to compare properly static pressure results with dynamic stress results. Graham et al. (1966) show that for germanium with a HEL of 44 kbar in the [111] orientation, a shock compression measurement of a phase transition at about 140 kbar corresponds to a dynamic pressure of about 120 kbar, in good agreement with static measurements of 120 to 125 kbar

(Minomura and Drickamer, 1962). Likewise, Larson (1967) showed that dynamic transition pressures for bismuth, corrected for strength-of-material effect and temperature, agree nicely with the value of 25.4 kbar obtained by static compression.

At high pressures, where strength-of-material effects are no longer important, it is expected that static and dynamic results should be in good agreement, if corrected for shock heating. In fact, dynamic results can be used to provide calibration points for static equipment in the hundreds of kilobars range. Drickamer (1970) presents a revised calibration for his high-pressure electrical resistance cell. Many o the calibration data were taken with aluminium or silver powder using the shock data of Rice *et al.* (1958), corrected to room temperature. The new calibration curve is lower than the older one, bringing the measured phase transition pressure points to lower pressures. For example, the new (α, ϵ) point for iron is listed at 110 to 115 kbar rather than at 133 kbar. The maximum pressure reached statically is about 450 kbar.

Decker (1965, 1966) calculated the equation of state of NaCl over a pressure range of 0 to 500 kbar for temperatures between 0 and 1500°C. He used the Mie–Gruneisen equation of state with Born–Mayer repulsion terms between first and second nearest neighbours. The calculated pressure–volume data agree with Bridgman's room-temperature measurement (Bridgman, 1940, 1945) in NaCl below 100 kbar to within 3 per cent and with high-pressure shock data (Al'tshuler *et al.*, 1961) to better than 2 per cent. More recent shock data (Fritz *et al.*, 1971), reduced to an isotherm, lies above Decker's pressure–volume isotherm. Fritz *et al.* point out that if Decker had chosen a bulk modulus equivalent to the sound speed they used, the two isotherms would be in essential agreement. The shock data are in agreement well within experimental error with the static data of Perez-Albuerne and Drickamer (1965). In view of the phase transformation in NaCl at about 300 kbar (Bassett *et al.*, 1968), Decker's equation may not be valid above 300 kbar.

In a recent paper, Vaidya and Kennedy (1970) make detailed comparisons of the compressions of a number of metals determined by three different experimental methods. The compression data obtained in a static, high-pressure piston-cylinder apparatus were found in agreement with the isothermal compressions calculated from shock-wave data up to 45 kbar. When no phase transition is present, ultrasonic velocity data can also be used to derive the isothermal equation of state. Agreement with the other two methods was generally fair.

The comparison of dynamic and static data is discussed in detail by Grover (1970), with special reference to the paper of Vaidya and Kennedy (1970). Grover finds that all three types of compression data are consistent up to 45 kbar.

Acknowledgments

The authors thank G. R. Abrahamson, T. W. Barbee, M. Cowperthwaite, P. S. DeCarli, M. J. Ginsberg, D. R. Grine, L. B. Hall, D. D. Keough, and L. Seaman for their helpful discussions and criticisms of the manuscript. Many authors kindly gave permission to use their published figures; individual credits appear in the Figure captions. Financial support was given by Stanford Research Institute.

References

Adadurov, G. A., Aliev, Z. G., Atovmyan, L. O., Bavina, T. V., Borod'ko, Y. G., Breusov, O. N., Dremin, A. N., Muranevich, A. K. and Pershin, S. V. (1967). *Sov. Phys.-Dokl.* **12**, 173–175.

Adler, D. and Brooks, H. (1965). *In* "Physics of Solids at High Pressures" (Eds C. T. Tomizuka and R. M. Emrick), pp. 567–570. Academic Press, New York and London.

Ahrens, T. J. (1966). *J. Appl. Phys.* **37**, 2532–2540.

Ahrens, T. J. (1972). *J. Appl. Phys.* **43**, 2443–2445.

Ahrens, T. J. and Gregson, V. G. (1964). *J. Geophys. Res.* **69**, 4839–4874.

Ahrens, T. J. and Ruderman, M. H. (1966). *J. Appl. Phys.* **37**, 4758–4765.

Ahrens, T. J., Gust, W. H. and Royce, E. B. (1968). *J. Appl. Phys.* **39**, 4610–4616.

Ahrens, T. J., Anderson, D. L. and Ringwood, A. E. (1969). *Rev. Geophys. Res.* **7**, 667–707.

Ainsworth, D. L. and Sullivan, B. R. (1967). U.S. Army Corps Engineering Technical Report 6–802, U. S. Army Engineers Waterways Experiment Station, Vicksburg, Mississippi.

Alder, B. J. and Christian, R. H. (1956). *Phys. Rev.* **104**, 550–551.

Alder, B. J. and Christian, R. H. (1961). *Phys. Rev. Lett.* **7**, 367–369.

Allison, F. E. (1965). *J. Appl. Phys.* **36**, 2111–2113.

Al'tshuler, L. V. (1965). *Sov. Phys.–Usp.* **8**, 52–91.

Al'tshuler, L. V. and Bakanova, A. A. (1969). *Sov. Phys.–Usp.* **11**, 678–689.

Al'tshuler, L. V., Bakanova, A. A. and Trunin, R. F. (1958a). *Sov. Phys.-Dokl.* **3**, 761–763.

Al'tshuler, L. V., Krupnikov, K. K. and Brazhnik, M. I. (1958b). *Sov. Phys.–JETP*, **7**, 614–619.

Al'tshuler, L. V., Kormer, S. B., Brazhnik, M. I., Vladimirov, L. A., Speranskaya, M. P. and Funtikov, A. I. (1960). *Sov. Phys.–JETP*, **11**, 766–775.

Al'tshuler, L. V., Kuleshova, L. V. and Pavlovskii, M. N. (1961). *Sov. Phys.–JETP*, **12**, 10–15.

Al'tshuler, L. V., Bakanova, A. A. and Dudoladov, I. P. (1966). *JETP Lett.* **3**, 315–317.

Al'tshuler, L. V., Pavlovskii, M. N. and Drakin, V. P. (1967). *Sov. Phys.–JETP*, **25**, 260–265.

Al'tshuler, L. V., Bakanova, A. A. and Dudoladov, I. P. (1968). *Sov. Phys.–JETP*, **26**, 1115–1120.

Anderson, G. D., Duvall, G. E., Erkman, J. O., Fowles, G. R. and Peltzer, C. P. (1966). Technical Report No. AFWL-TR-69-146, Stanford Research Institute, Contract AF29(601)-6429.

Anderson, G. D., Murri, W. J., Alverson, R. C. and Hanagud, S. V. (1967). Air Force Weapons Laboratory Technical Report AFWL-TR-67-24.

Andrews, D. J. (1973). *J. Phys. Chem. Solids*, **34**, 825–840.

Asay, J. R. and Gupta, Y. M. (1972). *J. Appl. Phys.* **43**, 2220–2223.

Asay, J. R., Fowles, G. R., Duvall, G. E., Miles, M. H. and Tinder, R. F. (1972a). *J. Appl. Phys.* **43**, 2132–2145.

Asay, J. R., Fowles, G. R. and Gupta, Y. (1972b). *J. Appl. Phys.* **43**, 744–746.

Backman, M. E. (1964). *J. Appl. Phys.* **35**, 2524–2533.

Bailey, L. and Lutze, A. (1972). Physics International Company Report. No. PIFR-246, Defense Nuclear Agency, Contract No. DASA01-70-C-0114.

Bakanova, A. A. and Dudoladov, I. P. (1967). *JETP Lett.* **5**, 265–267.

Bakanova, A. A., Dudoladov, I. P. and Sutulov, Yu. N. (1970). *Sov. Phys.–Solid State*, **11**, 1515–1518.

Balchan, A. S. (1962). *J. Appl. Phys.* **34**, 241–245.

Balchan, A. S. and Drickamer, H. G. (1961a). *J. Chem. Phys.* **34**, 1948–1949.

Balchan, A. S. and Drickamer, H. G. (1961b). *Rev. Sci. Instrum.* **32**, 308–313.

Bancroft, D. Peterson, E. L. and Minshall, F. S. (1956). *J. Appl. Phys.* **27**, 291–298.

Band, W. (1959). "Introduction to Mathematical Physics", p. 109, D. Van Nostrand, Princeton, New Jersey.

Band, W. and Anderson, G. D. (1962). *Amer. J. Phys.* **30**, 831–838.

Barbee, T., Seaman, L. and Crewdson, R. (1970). Final Report, Stanford Research Institute Contract F29601-68-C-0118, Air Force Weapons Laboratory, Kirtland AFB, Albuquerque, New Mexico.

Barbee, T. W., Seaman, L., Crewdson, R. and Curran, D. (1972). *J. Mater.* **7**, 393–401.

Barker, L. M. (1968). *In* "Behavior of Dense Media Under High Dynamic Pressures", pp. 483–506. Gordon and Breach, New York.

Barker, L. M. (1971a). *J. Compos. Mater.* **5**, 140–162.

Barker, L. M. (1971b). *Rev. Sci. Instrum.* **42**, 276–278.

Barker, L. M. (1972). *Exp. Mech.* **12**, 209–215.

Barker, L. M. and Hollenbach, R. E. (1964). *Rev. Sci. Instrum.* **35**, 742–746.

Barker, L. M. and Hollenbach, R. E. (1965). *Rev. Sci. Instrum.* **36**, 1617–1620.

Barker, L. M. and Hollenbach, R. E. (1970). *J. Appl. Phys.* **41**, 4208–4226.

Barker, L. M. and Hollenbach, R. E. (1972). *J. Appl. Phys.* **43**, 4669–4675.

Barker, L. M., Lundergan, C. D. and Herrmann, W. (1964). *J. Appl. Phys.* **35**, 1203–1212.

Barker, L. M., Butcher, B. M. and Karnes, C. H. (1966). *J. Appl. Phys.* **37**, 1989–1991.

Barsis, E., Williams, E. and Skoog, C. (1970). *J. Appl. Phys.* **41**, 5155–5162.

Bassett, W. A., Takahashi, T., Mao, H. and Weaver, J. S. (1968). *J. Appl. Phys.* **39**, 319–325.

Batsanov, S. S. and Lapshin, A. I. (1967). *Combust. Explos. Shock Waves (USSR)*, **3**, 271–275.

Baum, N., Shunk, R. and Bunker, R. (1973). Air Force Weapons Laboratory Technical Report AFWL-TR-73-24.

Belyakov, L. V., Valitskii, V. P., Zlatin, N. A. and Mochalov, S. M. (1967). *Sov. Phys.-Dokl.* **11**, 808–810.

Berg, U. I. (1963). *Ark. Fys.* **25**, 111–122.

Bergmann, O. R. and Barrington, J. (1966). *J. Amer. Ceram. Soc.* **49**, 502–507.

Bernstein, D. and Keough, D. D. (1962). Stanford Research Institute Formal Report Project 3713. Defense Atomic Support Agency, Washington, D.C., June 15.

Bernstein, D. and Keough, D. D. (1964). *J. Appl. Phys.* **35**, 1471–1474.

Bernstein, D., Godfrey, C., Klein, A. and Shimmin, W. (1968). *In* "Behavior of Dense Media Under High Dynamic Pressures", pp. 401–468. Gordon and Breach, New York.

Bland, D. R. (1965). *J. Inst. Math. Its Appl.* **1**, 56–75.

Boade, R. R. (1968a). *J. Appl. Phys.* **39**, 1609–1617.

Boade, R. R. (1968b). *J. Appl. Phys.* **39**, 5693–5702.

Boade, R. R. (1969). *J. Appl. Phys.* **40**, 3781–3785.

Boade, R. R. (1970a). *J. Appl. Phys.* **41**, 4542–4551.

Boade, R. R. (1970b). Report SC-DC-70-5052, Sandia Laboratories.

Bowden, H. G. and Kelly, P. M. (1967). *Acta Met.* **15**, 1489–1500.

Brazhnik, M. I., Krupnikov, K. K. and Krupnikova, V. P. (1962). *Sov. Phys.-JETP*, **15**, 470–476.

Breed, B. R. and Venable, D. (1968). *J. Appl. Phys.* **39**, 3222–3224.

Breed, B. R., Mader, C. L. and Venable, D. (1967). *J. Appl. Phys.* **38**, 3271–3275.

Bridgman, P. W. (1940). *Proc. Amer. Acad. Arts Sci.* **74**, 21–51.

Bridgman, P. W. (1942). *Proc. Amer. Acad. Arts Sci.* **74**, 399–424.

Bridgman, P. W. (1945). *Proc. Amer. Acad. Arts Sci.* **76**, 1–7.

Bridgman, P. W. (1950). *Proc. Roy. Soc., London*, **A203**, 1–17.

Bridgman, P. W. (1952). *Proc. Amer. Acad. Arts Sci.* **81**, 169–251.

Brillhart, D. C., DeAngelis, R. J., Preban, A. G., Cohen, J. B. and Gordon, P. (1967). *Trans. Met. Soc. AIME*, **239**, 836–843.

Bundy, F. P. (1965). *J. Appl. Phys.* **36**, 616–620.

Bundy, F. P. and Strong, H. M. (1962). *In* "Solid State Physics" (Eds F. Seitz and D. Turnbull), Vol. 13, pp. 81–146. Academic Press, New York and London.

Bundy, F. P. and Wentorf, R. H. (1963). *J. Chem. Phys.* **38**, 1144–1149.

Butcher, B. M. (1966). Report No. SC-DR-66-438, Sandia Corporation.

Butcher, B. M. (1968). *In* "Behavior of Dense Media Under High Dynamic Pressure", pp. 245–250. Gordon and Breach, New York.

Butcher, B. M. (1970). *In* Proceedings of the 17th Sagamore Army Materials Research Conference, published 1971: "Shock Waves and Mechanical Properties of Solids" (Eds J. J. Burke and V. Weiss), pp. 227–244. Syracuse University Press, Syracuse, N.Y.

Butcher, B. M. and Karnes, C. H. (1968). Report No. SC-RR-67-3040. Sandia Corporation.

Butcher, B. M. and Munson, D. E. (1968). *In* "Dislocation Dynamics" (Eds A. R. Rosenfield, G. T. Hahn, A. L. Bement, and R. I. Jaffee), pp. 591–607. McGraw-Hill, New York.

Butcher, B. M., Hicks, D. L. and Holdridge, D. B. (1972). Sandia Laboratories Report No. SC-RR-72-0627.

Buzhinskii, O. I. and Samylov, S. V. (1970). *Sov. Phys.–Solid State*, **11**, 2332–2336.

Cable, A. J. (1970). *In* "High-Velocity Impact Phenomena" (Ed. Ray Kinslow), pp. 1–21. Academic Press, New York and London.

Canfield, J. M. and Cutchen, J. T. (1966). *J. Appl. Phys.* **37**, 4581.

Carpenter, S. H. (1968). *Phil. Mag.* **17**, 855–857.

Carroll, M. M. and Holt, A. C. (1972). *J. Appl. Phys.* **48**, 1626–1636.

Champion, A. R. (1968). *Nature*, **220**, 157–158.

Champion, A. R. (1969). *J. Appl. Phys.* **40**, 3766–3771.

Champion, A. R. (1971). *J. Appl. Phys.* **42**, 5546–5550.

Champion, A. R. (1972). *J. Appl. Phys.* **43**, 2216–2220.

Champion, A. R. and Bartel, L. C. (1970). *Bull. Amer. Phys. Soc.* **15**, 1626.

Champion, A. R. and Benedick, W. B. (1968). *Rev. Sci. Instrum.* **39**, 377–378.

Chang, H. L. and Horie, Y. (1972). *J. Appl. Phys.* **43**, 3362–3366.

Chao, E. C. T. (1967). *Science*, **156**, 192–202.

Charest, J. A., Horne, D. E. and Jenrette, B. D. (1969). *Bull. Amer. Phys. Soc.* **14**, 1170.

Chen, P. J. and Gurtin, M. E. (1970). *Arch. Ration. Mech. Anal.* **36**, 33–46.

Chen, P. J., Graham, R. A. and Davison, L. (1972). *J. Appl. Phys.* **43**, 5021–5027.

Chou, P. C. and Wang, A. S. D. (1970). *J. Comp. Mater.* **4**, 444–461.

Christman, D. R. and Froula, N. H. (1970). *AIAA J.* **8**, 477–482.

Christou, A. (1970a). *Phil. Mag.* **21**, 203–205.

Christou, A. (1970b). *Phil. Mag.* **22**, 93–99.

Christou, A. (1972). *Phil. Mag.* **266**, 97–111.

Christou, A. and Brown, N. (1973). *Phil. Mag.* **27**, 281–296.

Clator, I. G. and Rose, M. F. (1967). *Brit. J. Appl. Phys.* **18**, 853–855.

Clifton, R. J. (1970a). *In* Proceedings of 17th Sagamore Army Materials Research Conference, published 1971: "Shock Waves and the Mechanical Properties of Solids" (Eds J. J. Burke and V. Weiss), pp. 73–120. Syracuse University Press, Syracuse, N.Y.

Clifton, R. J. (1970b). *J. Appl. Phys.* **41**, 5335–5337.

Cohen, J. B., Nelson, A. and DeAngelis, R. J. (1966). *Trans. Met. Soc. AIME*, **236**, 133–135.

Coleburn, N. L. and Forbes, J. W. (1968). *J. Chem. Phys.* **48**, 555–559.

Coleburn, N. L., Forbes, J. W. and Jones, H. D. (1972). *J. Appl. Phys.* **43**, 5007–5012.
Coleman, B. D. (1964). *Arch. Ration. Mech. Anal.* **17**, 1–46.
Coleman, B. D. and Gurtin, M. E. (1965a). *Arch. Ration. Mech. Anal.* **19**, 239–265.
Coleman, B. D. and Gurtin, M. E. (1965b). *Arch. Ration. Mech. Anal.* **19**, 266–298.
Coleman, B. D. and Gurtin, M. E. (1966). *Proc. Roy. Soc., London*, **A292**, 562–574.
Coleman, B. D., Gurtin, M. E. and Herrera, R. (1965). *Arch. Ration. Mech. Anal.* **19**, 1–19.
Coleman, B. D., Greenberg, J. M. and Gurtin, M. E. (1966). *Arch. Ration. Mech. Anal.* **22**, 333–354.
Conze, H., Crosnier, J. and Berard, C. (1968). *In* "Behavior of Dense Media Under High Dynamic Pressures", pp. 441–452. Gordon and Breach, New York.
Courant, R. and Friedrichs, K. (1948). "Supersonic Flow and Shock Waves". Interscience, New York.
Cowan, G. R. (1965). *Trans. Met. Soc. AIME*, **233**, 1120–1130.
Cowin, S. C. (1972). *J. Appl. Phys.* **43**, 2495–2497.
Cowperthwaite, M. (1968). *J. Franklin Inst.* **285**, 275–284.
Cowperthwaite, M. (1969). *J. Franklin Inst.* **287**, 379–387.
Cowperthwaite, M. (1972). *In* "Proceedings Fourteenth Symposium (International) on Combustion". *Combust. Inst.* 1231–1236.
Cowperthwaite, M. and Williams, R. F. (1971). *J. Appl. Phys.* **42**, 456–462.
Crosnier, J., Jacquesson, J. and Migault, A. (1965). *In* "Proceedings, Fourth Symposium on Detonation", pp. 627–638. Office of Naval Research, ACR-126, Washington, D.C.
Curran, D. R. (1961). *J. Appl. Phys.* **32**, 1811–1814.
Curran, D. R. (1963). *J. Appl. Phys.* **34**, 2677–2685.
Cutchen, J. T. (1966). *J. Appl. Phys.* **37**, 4745–4750.
Dapoigny, J., Kieffer, J. and Vodar, B. (1957). *C. R. Acad. Sci.* **245**, 1502–1505.
Darnell, A. J. and Libby, W. F. (1963). *Science*, **139**, 1301.
Davis, R. O. (1972). *Phys. Kondens Materie*, **15**, 230–236.
Davis, R. O., Jr. and Cottrell, M. M. (1971). *J. Compos. Mater.* **5**, 478–490.
Davis, R. O., Jr. and Wu, J. H. (1972). *J. Compos. Mater.* **6**, 126–135.
Deal, W. F. (1962). *In* "Modern Very High Pressure Techniques", (Ed. R. H. Wentorf, Jr.), pp. 200–227. Butterworths, Washington, D.C.
DeCarli, P. S. (1967a). *Bull. Amer. Phys. Soc.* **12**, 1127.
DeCarli, P. S. (1967b). *In* "Science and Technology of Industrial Diamonds" (Ed. J. Burk), Vol. 1, pp. 49–63. Industrial Diamond Information Bureau, London.
DeCarli, P. S. and Jamieson, J. C. (1961). *Science*, **133**, 1821–1822.
Decker, D. L. (1965). *J. Appl. Phys.* **36**, 157–161.
Decker, D. L. (1966). *J. Appl. Phys.* **37**, 5012–5014.
Dieter, G. E., Jr. (1961). "Mechanical Metallurgy", p. 254. McGraw-Hill, New York.
Doran, D. G. (1963). *In* "High Pressure Measurement" (Eds A. A. Giardini and E. C. Lloyd), pp. 59–86. Butterworths, London and Washington, D. C.
Doran, D. G. (1968). *J. Appl. Phys.* **39**, 40–47.
Doran, D. G. and Ahrens, T. J. (1963). Final Report, Stanford Research Institute Contract No. DA-04-200-ORD-1279.

Doran, D. G. and Linde, R. K. (1966). In "Solid State Physics", Vol. 19, pp. 229–290. Academic Press, New York and London.

Doran, D. G. and Murri, W. (1964). Bull. Amer. Phys. Soc. 9, 728

Dremin, A. N. and Adadurov, G. A. (1964). Sov. Phys.–Solid State, 6, 1379–1384.

Dremin, A. N. and Breusov, O. N. (1968). Russ. Chem. Rev. 37, 392–402.

Dremin, A. N. and Pershin, S. V. (1968). Combust. Explos. Shock Waves, 4, 112–115.

Dremin, A. N. and Shvedov, K. K. (1964). J. Appl. Mech. Tech. Phys. (USSR), 2, 154–159.

Dremin, A. N., Pershin, S. V. and Pogorelov, V. F. (1965). Combust. Explos. Shock Waves, 1, 1–4.

Drickamer, H. G. (1963). In "Solids Under Pressure" (Eds W. Paul and D. M. Warschauer), pp. 357–384. McGraw-Hill, New York.

Drickamer, H. G. (1970). Rev. Sci. Instrum. 41, 1667–1668.

Duff, R. E., Ross, M., Gust, W. H., Royce, E. B., Mitchell, A. C., Keeler, R. N. and Hoover, W. G. (1968). In "Behaviour of Dense Media Under High Dynamic Pressures", pp. 397–406. Gordon and Breach, New York.

Dugdale, J. S. and MacDonald, D. K. C. (1953). Phys. Rev. 89, 832–834.

Dulin, I. N., Al'tshuler, L. V., Vashchenko, V. Y. and Zubarev, V. N. (1969). Sov. Phys.–Solid State, 11, 1016–1020.

Duvall, G. E. (1962). In "Les Ondes de Detonation", p. 337. Editions de Centre National de la Recherche, Paris.

Duvall, G. E. (1971). In "Physics of High Energy Density" (Eds P. Caldirola and H. Knoepfel), pp. 7–50. Academic Press, New York and London.

Duvall, G. E. and Fowles, G. R. (1963). In "High Pressure Physics and Chemistry" (Ed. R. S. Bradley), Vol. 2, pp. 209–291. Academic Press, New York and London.

Duvall, G. E. and Taylor, S. M., Jr. (1971). J. Compos. Mater. 5, 130–139.

Duwez, P. (1935). Phys. Rev. 47, 494–501.

Eden, G. and Smith, C. P. M. (1970). In "Proceedings of Fifth Symposium (International) on Detonation", pp. 467–475. Office of Naval Research ACR-184, Arlington, Virginia.

Eden, G. and Wright, P. W. (1965). In "Proceedings, Fourth Symposium on Detonation", pp. 573–583. Office of Naval Research, ACR-126, Washington, D.C.

Egorov, L. A., Nitochkina, E. V. and Orekin, Yu. K. (1972). JETP Lett. 16, 4–6.

Eichelberger, R. J. and Hauver, G. E. (1962). In "Les ondes de Detonation", Editions de Centre National de la Recherche Scientifique, 15 Quai Anatole-France, Paris (VIIe).

Erkman, J. O. and Christensen, A. B. (1967). J. Appl. Phys. 38, 5395–5403.

Eshelby, J. D., Newey, C. W. A., Pratt, P. L. and Lidiard, A. B. (1958). Phil. Mag. 3, 75–89.

Falicov, L. M. (1965). In "Physics of Solids at High Pressures" (Eds C. T. Tomizuka and R. M. Emrick), pp. 30–45. Academic Press, New York and London.

Farber, J. and Ciolkosz, T. (1969). AFWL TR-69-18, Air Force Weapons Laboratory, Albuquerque, New Mexico.

Foltz, J. V. and Grace, F. I. (1969). J. Appl. Phys. 40, 4195–4199.

Fowles, G. R. (1960). J. Appl. Phys. 31, 655–661.

Fowles, G. R. (1961). J. Appl. Phys. 32, 1475–1487.

Fowles, G. R. (1967). *J. Geophys. Res.* **72**, 5729–5742.

Fowles, G. R. (1970) *J. Appl. Phys.* **41**, 2740–2741.

Fowles, G. R. (1973). Paper Presented at "Conference on Metallurgical Effects at High Strain Rates", Albuquerque, New Mexico, February. Proceeding to be published.

Fowles, G. R. and Williams, R. F. (1970). *J. Appl. Phys.* **41**, 360–363.

Frasier, J. T. and Karpov, B. G. (1965). *Exp. Mech.* **5**, 305–312.

Freeman, J. R. (1969). Sandia Laboratories Research Report SC-RR-69-55.

French, B. M. and Short, N. M. (ed.) (1968). "Shock Metamorphism of Natural Materials": Mono Book Corporation, Baltimore.

Fritz, J. N., Marsh, S. P., Carter, W. J. and McQueen, R. G. (1971). *In* "Accurate Characterization of The High-Pressure Environment" (Ed. E. C. Lloyd), pp. 201—208. National Bureau of Standards, Special Publication 326, Washington, D.C.

Fritzsche, J. (1965). *In* "Physics of Solids at High Pressures" (Eds C. T. Tomizuka and R. M. Emrick), pp. 184–195. Academic Press, New York and London.

Fuller, P. J. A. and Price, J. H. (1962). *Nature*, **193**, 262–263.

Fuller, P. J. A. and Price, J. H. (1964). *Brit. J. Appl. Phys.* **15**, 751–758.

Fuller, P. J. A. and Price, J. H. (1965). *In* "Proceedings Fourth Symposium on Detonation", pp. 290–294. Office of Naval Research, ACR-126, Washington, D.C.

Fuller, P. J. A. and Price, J. H. (1969). *Brit. J. Appl. Phys.* (*J. Phys. D.*), **2** 275–286.

Gaffney, E. S. (1972). *Trans. Amer. Geophys. Union*, **53**, 527.

Gaffney, E. S. and Ahrens, T. J. (1972). *Bull. Amer. Phys. Soc.* **17**, 1100.

Gager, W. B., Klein, M. J. and Jones, W. H. (1964). *Appl. Phys. Lett.* **5**, 131–132.

Garg, S. K. and Kirsch, J. W. (1971). *J. Compos. Mater.* **5**, 428–445.

Garg, S. K. and Kirsch, J. W. (1973). *J. Compos. Mater.* **7**, 277–285.

Geballe, T. H. and Hull, G. W. (1955). *Phys. Rev.* **98**, 940–947.

George, J. (1967). *Phil. Mag.* **15**, 497–506.

Gillis, P. P. (1970). *Scr. Met.* **4**, 533–538.

Gillis, P. P. and Gilman, J. J. (1965). *J. Appl. Phys.* **36**, 3370–3380 and 3380–3386.

Gillis, P. P., Gilman, J. J. and Taylor, J. W. (1969). *Phil. Mag.* **20**, 279–289.

Gillis, P. P., Hoge, K. G. and Wasley, R. J. (1971). *J. Appl. Phys.* **42**, 2145–2146.

Gilman, J. J. (1960). *Aust. J. Phys.* **13**, 327–346.

Gilman, J. J. (1965). *J. Appl. Phys.* **36**, 3195–3206.

Gilman, J. J. (1968). *Appl. Mech. Rev.* **21**, 767–783.

Gilson, V. A. (1967). *Appl. Opt.* **6**, 1217–1219.

Ginsberg, M. J. (1971). Stanford Research Institute Final Report, Defense Nuclear Agency, DNA 2742F.

Ginsberg, M. J. (1973). Stanford Research Institute Final Report (Draft), Defense Nuclear Agency Contract No. DNA001-72-C-0146.

Goranson, W., Bancroft, D., Burton, B. L., Belechar, T., Huston, E. E., Gittings, E. F. and Landeen, S. A. (1955). *J. Appl. Phys.* **26**, 1472–1479.

Goroff, I. and Kleinman, L. (1963). *Phys. Rev.* **132**, 1080–1084.

Gourley, L. E. and Gourley, M. F. (1968). *Bull. Amer. Phys. Soc.* **13**, 1660.

Grace, F. I. (1969). *J. Appl. Phys.* **40**, 2649–2653.

Grace, F. I. and Inman, M. C. (1970). *Metallography*, **3**, 89–98.

Grace, F. I., Inman, M. C. and Murr, L. E. (1968). *Brit. J. Appl. Phys.* (*J. Phys. D*), **1**, 1437–1444.

Grady, D. E. (1969). *Rev. Sci. Instrum.* **40**, 1399–1402.
Grady, D. E. (1972a). *Rev. Sci. Instrum.* **43**, 800–804.
Grady, D. E. (1972b). *J. Appl. Phys.* **43**, 1942–1948.
Grady, D. E. (1973). *J. Geophys. Res.* **78**, 1299–1307.
Grady, D. E., Duvall, G. E. and Royce, E. B. (1972). *J. Appl. Phys.* **43**, 1948–1955.
Graham, R. A. (1961). *Rev. Sci. Instrum.* **32**, 1308–1313.
Graham, R. A. (1962). *J. Appl. Phys.* **33**, 1755–1758.
Graham, R. A. (1968). *J. Appl. Phys.* **39**, 437–439.
Graham, R. A. (1972a). *Bull. Amer. Phys. Soc.* **17**, 1100.
Graham, R. A. (1972b). *Phys. Rev. B,* **12**, 4779–4792.
Graham, R. A. (1972c). *J. Acoust. Soc. Amer.* **51**, 1576–1581.
Graham, R. A. (1973). *Solid State Commun.* **12**, 503–505.
Graham, R. A. and Brooks, W. P. (1971). *J. Phys. Chem. Solids,* **32**, 2311–2329.
Graham, R. A. and Halpin, W. J. (1968). *J. Appl. Phys.* **39**, 5077–5082.
Graham, R. A. and Ingram, G. E. (1968). *Bull. Amer. Phys. Soc.* **13**, 1660.
Graham, R. A. and Ingram, G. E. (1969). *Bull. Amer. Phys. Soc.* **14**, 1163.
Graham, R. A. and Ingram, G. E. (1972). *J. Appl. Phys.* **43**, 826–835.
Graham, R. A., Neilson, F. W. and Benedick, W. B. (1965a). *J. Appl. Phys.* **36**, 1775–1783.
Graham, R. A., Jones, O. E. and Holland, J. R. (1965b). *J. Appl. Phys.* **36**, 3955–3956.
Graham, R. A., Jones, D. E. and Holland, J. R. (1966). *J. Phys. Chem. Solids,* **27**, 1519–1529.
Graham, R. A., Anderson, D. H. and Holland, J. R. (1967). *J. Appl. Phys.* **38**, 223–229.
Green, A. E. and Naghdi, P. M. (1965). *Arch. Ration. Mech. Anal.* **18**, 251–281.
Greenberg, J. M. (1967). *Arch. Ration. Mech. Anal.* **24**, 1–21.
Grover, R. (1970). *J. Phys. Chem. Solids,* **31**, 2347–2351.
Grover, R. and Rogers, F. J. (1973). *J. Phys. Chem. Solids,* **34**, 759–761.
Grüneisen, E. (1926). *In* "Handbuck der Physik", Vol. 10, p. 22, Verlag J. Springer, Berlin.
Gurtin, M. E. (1968). *Arch. Ration. Mech. Anal.* **28**, 40–50.
Gust, W. H. and Royce, E. B. (1970a). *J. Appl. Phys.* **41**, 2443–2446.
Gust, W. H. and Royce, E. B. (1970b). University California Lawrence Radiation Laboratory, Livermore, Calif. UCRL-50901.
Gust, W. H. and Royce, E. B. (1971a). *J. Appl. Phys.* **42**, 1897–1905.
Gust, W. H. and Royce, E. B. (1971b). *J. Appl. Phys.* **42**, 276–295.
Gust, W. H. and Royce, E. B. (1972). *J. Appl. Phys.* **43**, 4437–4442.
Gust, W. H., Holt, A. C. and Royce, E. B. (1973). *J. Appl. Phys.* **44**, 550–560.
Halpin, W. J. (1966). *J. Appl. Phys.* **37**, 153–163.
Halpin, W. J. (1968). *J. Appl. Phys.* **39**, 3821–3826.
Halpin, W. J. and Graham, R. A. (1965). *In* "Proceedings, Fourth Symposium on Detonation", pp. 222–232. Office of Naval Research, ACR-126, Washington D.C.
Halpin, W. J., Jones, O. E. and Graham, R. A. (1963). *In* "Dynamic Behavior of Materials", pp. 208–218. ASTM Special Technical Bulletin No. 336, Philadelphia, Pennsylvania.

Hamann, S. D. (1966). *In* "Advances in High Pressure Research" (Ed. R. S. Bradley), Vol. 1, pp. 85–144. Academic Press, New York and London.

Hanneman, R. E., Banus, M. D. and Gatos, H. C. (1964). *J. Phys. Chem. Solids*, **25**, 293–302.

Hanneman, R. E. and Strong, H. M. (1964). ASME Winter Annual Meeting, New York, ASMR Technical Paper 64-WA/PT-21.

Hauver, G. E. (1960). *In* "Third Symposium on Detonation", Vol. 1, pp. 241–252. Office of Naval Research Report ACR-52.

Hauver, G. E. (1965). *J. Appl. Phys.* **36**, 2133–2118.

Hearst, J. R., Gessaman, L. B. and Power, D. V. (1964). *J. Appl. Phys.* **35**, 2145–2150.

Hearst, J. R., Irani, G. B. and Gessaman, L. B. (1965). *J. Appl. Phys.* **36**, 3440–3444.

Heckel, R. W. and Youngblood, J. L. (1968). *J. Amer. Ceram. Soc.* **51**, 398–401.

Herring, C. (1954). *Phys. Rev.* **96**, 1163–1187.

Herrmann, W. (1969a). *J. Appl. Phys.* **40**, 2490–2499.

Herrmann, W. (1969b). *In* "Wave Propagation in Solids" (Ed. J. Miklowitz), ASME.

Herrmann, W. and Nunziato, J. W. (1973). *In* "Dynamic Behavior of Materials Under Impulsive Loading" (Ed. P. C. Chou), Chapter 5. To be published.

Herrmann, W., Lawrence, R. J. and Mason, D. S. (1969). *In* "Proceedings of a Meeting on Radiation Effects", Vol. 1. Air Force Weapons Laboratory, Albuquerque, New Mexico, Dec. 3–4.

Herrmann, W., Lawrence, R. J. and Mason, D. S. (1970a). Sandia Laboratories Report SC-RR-70-741.

Herrmann, W., Hicks, D. L. and Young, E. G. (1970b). *In* Proceedings of the 17th Sagamore Army Materials Research Conference, published 1971: "Shock Waves and the Mechanical Properties of Solids" (Eds J. J. Burke and V. Weiss), pp. 23–63. Syracuse University Press, Syracuse, N.Y.

Heyda, J. F. (1969). *In* "Behavior of Dense Media Under High Dynamic Pressures", pp. 227–232. Gordon and Breach, New York.

Higgens, G. T. (1971). *Met. Trans.* **2**, 1277–1282.

Hofmann, R., Andrews, D. J. and Maxwell, D. E. (1968). *J. Appl. Phys.* **39**, 4555–4562.

Holland, J. R. (1967). *Acta Met.* **15**, 691–699.

Hopkins, A. K. (1970). Report AFWL-TR-70-158, Air Force Materials Laboratory, Wright-Patterson Air Force Base.

Horiguchi, Y. (1966). *Naturwiss.* **53**, 476.

Horie, Y. (1967). *J. Phys. Chem. Solids*, **28**, 1569–1577.

Huang, Y. K. (1970). *J. Chem. Phys.* **53**, 571–575.

Hughes, D. S., Gourley, L. E. and Gourley, M. F. (1961). *J. Appl. Phys.* **32**, 624–629.

Ingram, G. E. (1965). *Rev. Sci. Instrum.* **36**, 458–460.

Ingram, G. E. and Graham, R. A. (1970). *In* "Proceedings of the Fifth Symposium on Detonation", pp. 369–386. Office of Naval Research, ACR-184, Washington, D.C.

Ivanov, A. G., Novikov, S. A. and Sinitsyn, V. A. (1963). *Sov. Phys.–Solid State*, **5**, 196–202.

Ivanov, A. G., Mineev, V. N., Novitskii, E. Z., Yanov, V. A. and Bezrukov, G. I. (1965). *JETP Lett.* **2**, 223–224.

Ivanov, A. E., Novitskii, E. Z., Mineev, V. N., Lisitsyn, Yu. V., Tyunyaev, Yu. N. and Bezrukov, G. I. (1968a). *Sov. Phys.–JETP*, **26**, 28–32.

Ivanov, A. G., Lisitsyn, Yu. V. and Novitskii, E. Z. (1968b). *Sov. Phys.–JETP*, **27**, 153–155.

Ivanov, A. G., Mineev, V. N., Nivitskii, E. Z., Lisitsyn, Yu. V. and Tyunyaev, Yu. N. (1968c). *JETP Lett*, **7**, 147–149.

Jacobs, S. J. and Edwards, D. J. (1970). *In* "Proceedings of the Fifth Symposium on Detonation", pp. 413–426. Office of Naval Research, ACR-184, Washington, D.C.

Jacquesson, J. (1962). *In* "Les Ondes de Detonation", pp. 415–422. C.N.R.S., Paris.

Jacquesson, J., Romain, J. P., Hallouin, M. and Desoyer, J. C. (1970). *In* "Proceedings of the Fifth Symposium on Detonation", pp. 403–412. Office of Naval Research, ACR-184, Washington, D.C.

Jajosky, B. C., Jr. and Ferman, M. A. (1969). AFWC-TR-69-101, Air Force Weapons Laboratory, Kirtland AFB, Albuquerque, New Mexico.

Johnson, J. N. (1968). SC-RR-68-151, Sandia Corporation.

Johnson, J. N. (1969). *J. Appl. Phys.* **40**, 2287–2293.

Johnson, J. N. (1970). *Bull. Amer. Phys. Soc.* **15**, 1606.

Johnson, J. N. (1971). *J. Appl. Phys.* **42**, 5522–5530.

Johnson, J. N. (1972a). Sandia Laboratories Research Report, SC-RR-72 0626.

Johnson, J. N. (1972b). *J. Appl. Phys.* **43**, 2074–2082.

Johnson, J. N. and Band, W. (1967). *J. Appl. Phys.* **38**, 1578–1585.

Johnson, J. N. and Barker, L. M. (1969). *J. Appl. Phys.* **40**, 4321–4334.

Johnson, J. N. and Rohde, R. W. (1971). *J. Appl. Phys.* **42**, 4171–4182.

Johnson, J. N., Jones, O. T. and Michaels, T. E. (1970). *J. Appl. Phys.* **41**, 2330–2339.

Johnson, J. O. and Wackerle, J. (1967). *In* "Behaviour of Dense Media Under High Dynamic Pressure", pp. 217–226. Gordon and Breach, New York.

Johnson, P. M. and Burgess, T. J. (1968). *Rev. Sci. Instrum.* **39**, 1100–1103.

Johnson, Q. and Mitchell, A. C. (1972). *Phys. Rev. Lett.* **29**, 1369–1371.

Johnson, Q., Mitchell, A., Keeler, R. N. and Evans, L. (1970). *Phys. Rev. Lett.* **25**, 1099–1101.

Johnson, Q., Mitchell, A. C. and Evans, L. (1971). *Rev. Sci. Instrum.* **42**, 999–1001.

Johnston, W. G. (1962). *J. Appl. Phys.* **33**, 2716–2730.

Jones, A. H., Isbell, W. M. and Maiden, C. J. (1966). *J. Appl. Phys.* **37**, 3493–3499.

Jones, E. D. (1971). *Appl. Phys. Lett.* **18**, 33–35.

Jones, G. A. and Halpin, W. J. (1968). *Rev. Sci. Instrum.* **39**, 258–259.

Jones, O. E. (1967). *Rev. Sci. Instrum.* **38**, 253–256.

Jones, O. E. and Graham, R. A. (1971). *In* "Accurate Characterization of the High-Pressure Environment" (Ed. E. C. Lloyd), pp. 229–237. NBS Special Publication 326, Washington, D.C.

Jones, O. E. and Holland, J. R. (1964). *J. Appl. Phys.* **35**, 1771–1773.

Jones, O. E. and Holland, J. R. (1968). *Acta Met.* **16**, 1037–1045.

Jones, O. E. and Mote, J. D. (1969). *J. Appl. Phys.* **40**, 4920–4928.

Jones, O. E., Neilson, F. W. and Benedick, W. B. (1962). *J. Appl. Phys.* **33**, 3224–3232.

Kaechele, L. E. and Tetelman, A. S. (1969). *Acta Met.* **17**, 463–475.

Karnes, C. H. (1968). In "Mechanical Behavior of Materials Under Dynamic Loads" (Ed. U. S. Lindholm), pp. 270–203. Springer-Verlag, New York.

Keeler, R. N. and Mitchell, A. D. (1969). Solid State Commun. 7, 271–274.

Keeler, R. N. and Royce, E. G. (1971). In "Physics of High Energy Density" (Eds P. Caldirola and H. Knoepfel), pp. 51–150. Academic Press, New York and London.

Kelly, J. M. (1970). Arch. Mech. Stosowanej, 1, 22, 93–110.

Kelly, J. M. and Gillis, P. P. (1967). J. Appl. Phys. 38, 4044–4046.

Kelly, J. M. and Gillis, P. P. (1970). J. Appl. Mech. 163–170.

Kennedy, J. D. (1968). In "Behavior of Dense Media Under High Dynamic Pressures", pp. 407–418. Gordon and Breach, New York.

Kennedy, J. D. and Benedick, W. B. (1966). J. Phys. Chem. Solids, 27, 125–127.

Kennedy, J. D. and Benedick, W. B. (1967). Solid State Commun. 5, 53–55.

Keough, D. D. (1963). Stanford Research Institute Final Report, Project 3713. Defense Atomic Support Agency, Washington, D.C. DASA No. 1414.

Keough, D. D. (1964). Stanford Research Institute Final Report, Defense Atomic Support Agency, Contract No. DA-49-146-XZ-096.

Keough, D. D. (1968). Stanford Research Institute Technical Report AFWL-TR-68-57.

Keough, D. D. (1970). Stanford Research Institute Final Report, Defense Atomic Support Agency, Contract No. DASA-69-C-0014.

Keough, D. D. and Wong, J. Y. (1970). J. Appl. Phys. 41, 3508–3515.

Keough, D. D., Williams, R. F. and Bernstein, D. (1964). In "Symposium on High Pressure Technology" (Eds Lloyd and Giardini). Presented at Winter Annual Meeting of the American Society of Mechanical Engineers (November 29–December 4). American Society of Mechanical Engineers, United Engineering Center, New York.

Keough, D. D., Smith, C. W. and Cowperthwaite, M. (1971). Stanford Research Institute Interim Report, Project 8567, Defense Atomic Support Agency, Contract No. DASA01-70-C-0098.

Kirillov, G. A., Kormer, S. B. and Sinitsyn, M. V. (1968). JETP Lett. 1, 290–291.

Klein, M. J. and Edington, J. W. (1966). Phil. Mag. 14, 21–29.

Klein, M. J. and Rudman, P. S. (1966). Phil. Mag. 14, 1199–1206.

Knopoff, L. and Shapiro, J. N. (1969). J. Geophys. Res. 74, 1439–1450.

Kohn, W. (1965). In "Physics of Solids at High Pressures" (Eds C. T. Tomizuka and R. M. Emrick), pp. 561–566. Academic Press, New York and London.

Kompaneets, A. S., Romanova, V. I. and Yampol'skii, P. A. (1972). JETP Lett. 16, 183–185.

Kormer, S. B. (1968) Sov. Phys.-Uspekhi, 11, 229–254.

Kormer, S. B., Funtikov, A. I., Urlin, V. D. and Kolesnikova, A. N. (1962). Sov. Phys.–JETP, 15, 477–488.

Kormer, S. B., Sinitsyn, M. V., Kirillov, G. A. and Urlin, V. D. (1965). Sov. Phys.– JETP, 21, 689–700.

Kormer, S. B., Sinitsyn, M. V., Kirillov, G. A. and Popova, L. T. (1966a). Sov. Phys.– JETP, 22, 97–105.

Kormer, S. B., Yushko, K. B. and Krishkevich, G. V. (1966b). JETP Lett. 3, 39–42.

Kormer, S. B., Yushko, K. B. and Krishkevich, G. V. (1968). Sov. Phys.–JETP, 6, 879–881.

Kormer, S. B., Sinitsyn, M. V. and Kuryapin, A. I. (1969). *Sov. Phys.–JETP*, **28**, 852–854.

Kressel, H. and Brown, N. (1967). *J. Appl. Phys.* **38**, 1618–1625.

Krupnikov, K. K., Brazhnik, M. I. and Krupnikova, V. P. (1962). *Sov. Phys.–JETP*, **15**, 470–476.

Kuleshova, L. V. (1969). *Sov. Phys.–Solid State*, **11**, 886–890.

Kusubov, A. S. and van Thiel, M. (1969). *J. Appl. Phys.* **40**, 3776–3780.

Larson, D. B. (1965). *In* "Physics of Solids at High Pressures" (Eds C. T. Tomizuka and R. M. Emrick), p. 459. Academic Press, New York and London.

Larson, D. B. (1967). *J. Appl. Phys.* **38**, 1541–1546.

Latham, G., Ewing, M., Dorman, J., Press, F., Toksoz, N., Sutton, G., Meissner, R., Dunnebier, F., Nakamura, Y., Kovach, R. and Yates, M. (1970). *Science*, **170**, 620–626.

Lee, E. H. (1966). *In* "Proceedings of the Fifth U.S. National Congress of Applied Mechanics", pp. 405–420. ASME.

Lee, E. H. and Liu, D. T. (1967). *J. Appl. Phys.* **38**, 19–27.

Lee, E. H. and Wierzbicki, T. (1967). *J. Appl. Mech.* **34**, 931–936.

Lee, L. M. (1967). SC-RR-67-2947, Sandia Laboratory, December.

Lee, L. M. (1972). Sandia Laboratories Report No. SC-RR-72-0814.

Liebermann, P. (1967). Task Two Final Report, 11TR1-578P21-22, Project T6042, Illinois Institute of Technology Research Institute (October).

Lighthill, M. J. (1956). *In* "Surveys in Mechanics" (Eds G. K. Batchelor and P. M. Davies), pp. 250–351. Cambridge University Press, London.

Linde, R. K. and DeCarli, P. S. (1969). *J. Chem. Phys.* **50**, 319–325.

Linde, R. K. and Doran, D. G. (1966). *Nature*, **212**, 27–29.

Linde, R. K. and Schmidt, D. N. (1966a). *J. Appl. Phys.* **37**, 3259–3271.

Linde, R. K. and Schmidt, D. N. (1966b). *Rev. Sci. Instrum.* **37**, 1–7.

Linde, R. K., Murri, W. J. and Doran, D. G. (1966). *J. Appl. Phys.* **37**, 2527–2532.

Linde, R. K., Seaman, L. and Schmidt, D. N. (1972). *J. Appl. Phys.* **43**, 3367–3375.

Loree, T. R., Minshall, F. S. and Stirpe, D. (1965). Los Alamos Scientific Laboratory, LA-3215.

Loree, T. R., Fowler, C. M., Zukas, E. G. and Minshall, F. S. (1966a). *J. Appl. Phys.* **37**, 1918–1927.

Loree, T. R., Warnes, R. H., Zukas, E. G. and Fowler, C. M. (1966b). *Science*, **153**, 1277–1278.

Lundergan, C. D. and Drumheller, D. S. (1970). *In* Proceedings of the 17th Sagamore Army Materials Research Conference, published 1971: "Shock Waves and the Mechanical Properties of Solids" (Eds J. J. Burke and V. Weiss), pp. 141–154. Syracuse University Press, Syracuse, N.Y.

Lundergan, C. D. and Drumheller, D. S. (1971). *J. Appl. Phys.* **42**, 669–679.

Lundergan, C. D. and Drumheller, D. S. (1972). *In* "Dynamics of Composite Material" (Ed. E. H. Lee). Presented at 1972 Joint National and Western Applied Mechanics Conference, ASME (June 26–28). La Jolla, California, USA.

Lutze, A. (1971). Physics International Company Report, No. PIFR-341, Sandia Corporation, Contract No. 20-7277.

Lyle, J. W. (1968). University of California, Lawrence Livermore Laboratory Report. UCRL No. 71384; *Bull. Amer. Phys. Soc.* **13**, 1678.

Lyle, J. W. and Schriever, R. L. (1969). *J. Appl. Phys.* **40**, 4663–4664.

Lysne, P. C. (1968). *Rev. Sci. Instrum.* **39**, 754–755.

Lysne, P. C. (1970a). *J. Geophys. Res.* **75**, 4375–4386.

Lysne, P. C. (1970b). *J. Appl. Phys.* **41**, 351–360.

Lysne, P. C. (1972). *J. Appl. Phys.* **43**, 425–431.

Lysne, P. C. (1973). *J. Appl. Phys.* **44**, 577–582.

Lysne, P. C. and Halpin, W. J. (1968). *J. Appl. Phys.* **39**, 5488–5495.

Lysne, P. C., Boade, R. R., Percival, C. M. and Jones, O. E. (1969). *J. Appl. Phys.* **40**, 3786–3795.

Lysne, P. C., Graham, R. A., Bartel, L. C. and Samara, G. A. (1971). Sandia Laboratories Technical Memorandum, SC-TM-710907.

Ma, C. H. (1972). *Phys. Rev. B*, **5**, 2826–2829.

Ma, C. H., Mitchell, D. W. and Daga, R. (1969). *J. Appl. Phys.* **40**, 3499–3504.

Mahajan, S. (1969). *Phil. Mag.* **19**, 199–204.

Malvern, L. E. (1951). *Quart. Appl. Math.* **8**, 405–411.

Marsh, D. J. (1970). *Phil. Mag.* **21**, 95–108.

McQueen, R. G. (1964). *In* "Metallurgy at High Pressures and Temperatures" (Eds K. A. Gschneidner, Jr., M. T. Hepworth, and N. A. D. Parlee), pp. 44–132. Gordon and Breach, New York and London.

McQueen, R. G. and Marsh, S. P. (1960). *J. Appl. Phys.* **31**, 1253–1269.

McQueen, R. G. and Marsh, S. P. (1968). *In* "Behavior of Dense Media under High Dynamic Pressures", pp. 207–216. Gordon and Breach, New York and London.

McQueen, R. G., Marsh, S. P., Taylor, J. W., Fritz, J. N. and Carter, W. J. (1970). *In* "High-Velocity Impact Phenomena" (Ed. Ray Kinslow), pp. 293–417. Academic Press, New York and London.

Migault, A. (1970). Thesis submitted to Faculty of Sciences, University of Poitier.

Migault, A. and Jacquesson, J. (1968). *In* "Behavior of Dense Media Under High Dynamic Pressures", pp. 431–440. Gordon and Breach, New York.

Mikkola, D. E. and Cohen, J. B. (1966). *Acta Met.* **14**, 105–122.

Mineev, V. N., Ivanov, A. G., Novitskii, E. Z., Tyunyaev, Yu. N. and Lisitsyn, Yu. V. (1967). *JETP Lett.* **5**, 244–246.

Mineev, V. N., Tyunyaev, Yu. N., Ivanov, A. G., Novitskii, E. Z. and Lisitsyn, Yu. V. (1968). *Sov. Phys.–JETP*, **26**, 728–731.

Mineev, V. N., Ivanov, A. G., Lisitsyn, Yu. V., Novitskii, E. Z. and Tyunyaev, Yu. N. (1971). *Sov. Phys.–JETP*, **32**, 592–598.

Mineev, V. N., Ivanov, A. G., Lisitsyn, Yu. V., Novitskii, E. Z. and Tyunyaev, Yu. N. (1972). *Sov. Phys.–JETP*, **34**, 131–135.

Minomura, S. and Drickamer, H. G. (1962). *J. Phys. Chem. Solids*, **23**, 451–456.

Mitchell, A. C. and Keeler, R. N. (1967). *Bull. Amer. Phys. Soc.* **12**, 1128.

Mitchell, A. C., Johnson, Q. and Evans, L. (1973). *Rev. Sci. Instrum.* **44**, 597–599.

Morland, L. W. (1959). *Phil. Trans. Roy. Soc., London*, **A251**, 341–382.

Munson, D. E. and Barker, L. M. (1966). *J. Appl. Phys.* **37**, 1652–1660.

Munson, D. E. and Schuler, K. W. (1971). *J. Compos. Mater.* **5**, 286–304.

Murnaghan, F. D. (1937). *Amer. J. Math.* **59**, 235.

Murr, L. E. and Foltz, J. V. (1969). *J. Appl. Phys.* **40**, 3796–3802.

Murr, L. E. and Grace, F. I. (1969). *Exp. Mech.* **9**, 145–155.

Murr, L. E. and Rose, M. F. (1968). *Phil. Mag.* **18**, 281–295.

Murr, L. E. and Vydyanath, H. R. (1970a). *Acta Met.* **18**, 1047–1052.

Murr, L. E. and Vydyanath, H. R. (1970b). *Scr. Met.* **4**, 183–188.

Murr, L. E., Vydyanath, H. R. and Foltz, J. V. (1970). *Metallurgical Transactions*, **1**, 3215–3223.

Murr, L. E., Foltz, J. V. and Altman, F. D. (1971). *Phil. Mag.* **23**, 1011–1028.

Murri, W. J. (1972). Stanford Research Institute Final Report, Project 8909, U.S. Atomic Energy Commission, Contract No. AT(04-3)-115.

Murri, W. J. and Anderson, G. D. (1970). *J. Appl. Phys.* **41**, 3521–3525.

Murri, W. J. and Doran, D. G. (1965). Stanford Research Institute Project 4100, Contract No. DA-04-200-ORD-1279, Final Report.

Murri, W. J., Petersen, C. F. and Smith, C. W. (1969). *Bull. Amer. Phys. Soc.* **14**, 1155.

Naumann, R. J. (1971). *J. Appl. Phys.* **42**, 4945–4954.

Neilson, F. W. and Benedick, W. B. (1960). *Bull. Amer. Phys. Soc.* **5**, 511.

Neilson, F. W., Benedick, W. B., Brooks, W. P., Graham, R. A. and Anderson, G. W. (1962). *In* "Les Ondes de Detonation", pp. 391–414. Editions du Centre National de la Recherche Scientifique, Paris.

Ng, R. and Butcher, B. M. (1972). Sandia Laboratories Report No. SC-RR-71790.

Nomura, Y. and Tobisawa, S. (1965). *Appl. Phys. Lett.* **7**, 126–127.

Novikov, S. A., Sinitsyn, V. A., Ivanov, A. G. and Vasilyev, L. V. (1966). *Phys. Met. Metallogr.* **21**, 135–144.

Novitskii, E. Z., Kleschkvnikov, O. A., Mineev, V. N. and Ivanov, A. G. (1973). *Fiz. Tverd. Tela*, **15**, 310.

O'Brien, J. F. and Wasley, R. J. (1966). *Rev. Sci. Instrum.* **37**, 531–532.

Orava, R. N. (1973). *Mater. Sci. Eng.* **11**, 177–180.

Orowan, E. (1959). *In* "Internal Stresses and Fatigue in Metals". Elsevier, New York.

Palmer, E. P. and Turner, G. H. (1964). *J. Appl. Phys.* **35**, 3055–3056.

Pastine, D. J., Lombardi, M., Chatterjee, A. and Tchen, W. (1970). *J. Appl. Phys.* **41**, 3144–3147.

Paul, W. and Warshauer, D. M. (1963). *In* "Solids Under Pressure" (Eds W. Paul and D. M. Warshauer), pp. 179–249. McGraw-Hill, New York.

Pavlovskii, M. N. (1968). *Sov. Phys.–Solid State*, **9**, 2514–2518.

Pavlovskii, M. N. and Drakin, V. P. (1966). *JETP Lett.* **4**, 169–172.

Peck, J. C. (1970). *In* Proceedings of the 7th Sagamore Army Materials Research Conference, published 1971: "Shock Waves and the Mechanical Properties of Solids" (Eds J. J. Burke and V. Weiss), pp. 155–184. Syracuse University Press, Syracuse, N.Y.

Peck, J. C. (1972). *In* "Dynamics of Composite Material" (Ed. E. H. Lee). Presented at 1972 Joint National and Western Applied Mechanic Conference, ASME, (June 26–28), La Jolla, California, USA.

Perez-Albuerne, E. A. and Drickamer, H. G. (1965). *J. Chem. Phys.* **43**, 1381–1387.

Perls, T. A. (1952). *J. Appl. Phys.* **23**, 674–680.

Perry, F. C. (1970). *Appl. Phys. Lett.* **17**, 478–481.

Petersen, C. F. (1969). Ph.D. thesis, Stanford University, and Technical Report, Stanford Research Institute.

Petersen, C. F., Murri, W. J. and Cowperthwaite, M. (1970). *J. Geophys. Res.* **75**, 2063–2072.

Peyre, C., Pujol, J. and Thouvenin, J. (1965). *In* "Proceedings, Fourth Detonation Symposium", pp. 566–572. Office of Naval Research, ACR-126, Washington, D.C.

Plyukhin, V. S. and Kologrivov, V. N. (1962). *Zh. Prikl. Mekh. Tekh. Fiz.* No. **5**, 175–176. Translated in FTD-TT-63-1057 (1963).

Prindle, H. (1968). D2-125019-1, The Boeing Company (February).

Rayleigh, Lord (1911). *Proc. Roy. Soc., London,* **A84**, 247–284.

Rempel, J. R., Schmidt, D. N., Erkman, J. O. and Isbell, W. M. (1966). Technical Report WL-TR-64-119, Stanford Research Institute, Contract AF29(601)-6040 (February).

Rice, M. H., McQueen, R. G. and Walsh, J. M. (1958). *Solid State Phys.* **6**, 1–63.

Rohde, R. W. (1969a). *J. Appl. Phys.* **40**, 2988–2993.

Rohde, R. W. (1969b). *Acta Met.* **17**, 353–363.

Rohde, R. W. (1970). *Acta Met.* **18**, 903–913.

Rohde, R. W. and Jones, O. E. (1968). *Rev. Sci. Instrum.* **39**, 313–316.

Rohde, R. W. and Towne, T. L. (1971). *J. Appl. Phys.* **42**, 878–880.

Rohde, R. W., Holland, J. R. and Graham, R. A. (1968). *Trans. Met. Soc. AIME,* **242**, 2017–20191

Rose, M. F. and Berger, T. L. (1968). *Phil. Mag.* **17**, 1121–1133.

Rose, M. F. and Grace, F. I. (1967). *Brit. J. Appl. Phys.* **18**, 671–674.

Rose, M. F. and Inman, M. C. (1969). *Phil. Mag.* **19**, 925–930.

Rose, M. F., Berger, T. L. and Inman, M. C. (1967). *Trans. Met. Soc. AIME,* **239**, 1998–1999.

Rose, M. F., Villere, M. P. and Berger, T. L. (1969). *Phil. Mag.* **19**, 39–45.

Royce, E. B. (1966). *J. Appl. Phys.* **37**, 4066–4070.

Royce, E. B. (1967). *Phys. Rev.* **164**, 929–943.

Royce, E. B. (1968). *In* "Behavior of Dense Media Under High Dynamic Pressures", pp. 419–430. Gordon and Breach, New York.

Russell, T. L., Wood, D. S. and Clark, D. S. (1961). *Acta Met.* **9**, 1054–1063.

Samara, G. A. and Drickamer, H. G. (1962). *J. Phys. Chem. Solids,* **23**, 457–461.

Schall, R. (1958). *Explosivstoffe,* **6**, 120–124.

Schuler, K. W. (1970). *Mech. Phys. Solids,* **18**, 277–292.

Schuler, K. W. and Walsh, E. K. (1970). In preparation.

Seaman, L. (1970a). *In* Proceedings of the 17th Sagamore Army Materials Research Conference, published 1971: "Shock Waves and the Mechanical Properties of Solids" (Eds J. J. Burke and V. Weiss), pp. 245–261. Syracuse University Press, Syracuse, N.Y.

Seaman, L. (1970b). *In* "Principles of Explosive Behavior", Chapter 15. Brinkley and Gordon. To be published.

Seaman, L. (1970c). Final Report of Contract F29601-70-C-001 (January 2), prepared for Air Force Weapons Laboratory, Kirtland AFB, Albuquerque, New Mexico.

Seaman, L. and Linde, R. K. (1969). Technical Report No. AFWL-TR-68-143 (May), Vol. 1, Stanford Research Institute.

Seaman, L., Williams, R. F., Rosenberg, J. T., Erlich, D. C. and Linde, R. K. (1969). Technical Report No. AFWL-TR-69-96 (November), submitted to Air Force Weapons Laboratory, Kirtland AFB, Albuquerque, New Mexico.

Seaman, L., Barbee, T. W. and Curran, D. R. (1971). Stanford Research Institute Final Report, Air Force Weapons Laboratory, Contract F29601-70-C-0070.

Seay, G. E., Graham, R. A., Wayne, R. C. and Wright, L. D. (1967). *Bull. Amer. Phys. Soc.* **12**, 1129.

Shaner, J. W. and Royce, E. B. (1968). *J. Appl. Phys.* **39**, 492–493.

Shapiro, J. N. and Knopoff, L. (1969). *J. Geophys. Res.* **74**, 1435–1438.

Shockey, D. A., Seaman, L. and Curran, D. R. (1973). Stanford Research Institute Final Report, Air Force Weapons Laboratory, Contract No. F29601-70-C-0070.

Skidmore, I. C. (1967). *Appl. Mater. Res.* **4**, 131.

Slater, J. C. (1939). "Introduction to Chemical Physics". McGraw-Hill, New York.

Smith, J. H. (1963). *In* "Symposium on Dynamic Behavior of Materials", pp. 264–281. ASTM Special Technical Publication No. 336, Philadelphia.

Stevens, A. L. and Jones, O. E. (1972). *J. Appl. Mech. Trans. ASME*, **39**, 359–366.

Styris, D. L. and Duvall, G. E. (1970). *High Temp.–High Pres.* **2**, 477–499.

Swenson, C. A. (1960). *In* "Solid State Physics" (Eds F. Seitz and D. Turnbull), Vol. 17, pp. 41–147. Academic Press, New York and London.

Taborsky, L. (1969). *Phys. Status. Solidi.* **35**, K5–K8.

Taylor, G. I. (1911). *Proc. Roy. Soc., London*, **A84**, 371–377.

Taylor, J. W. (1963). *J. Appl. Phys.* **34**, 2727–2731.

Taylor, J. W. (1965). *J. Appl. Phys.* **36**, 3146–3150.

Taylor, J. W. (1968). *In* "Dislocation Dynamics" (Eds A. R. Rosenfield, G. T. Hahn, A. L. Bement, Jr. and R. I. Jaffe), pp. 573–589. McGraw-Hill, New York.

Taylor, J. W. and Rice, M. (1963). *J. Appl. Phys.* **34**, 364–371.

Thouvenin, J. (1965). *In* "Proceedings, Fourth Symposium on Detonation", pp. 258–265. Office of Naval Research, ACR-126, Washington, D.C.

Thurnborg, S., Ingram, G. E. and Graham, R. A. (1964). *Rev. Sci. Instrum.* **35**, 11–14.

Torvik, P. J. (1970). *J. Compos. Mater.* **4**, 296–309; AFIT-TR-70-3, Air Force Institute of Technology, Wright-Patterson AFB, Ohio.

Towne, T. L. (1970). *Rev. Sci. Instrum.* **41**, 290–291.

Trueb, L. F. (1968). *J. Appl. Phys.* **39**, 4707–4716.

Trueb, L. F. (1969). *J. Appl. Phys.* **40**, 2976–2987.

Trueb, L. F. (1970). *J. Appl. Phys.* **41**, 5029–5030.

Trueb, L. F. (1971). *J. Appl. Phys.* **42**, 503–510.

Truesdell, C. and Noll, W. (1965). "Handbuch der Physik", Vol. III, p. 3. Springer-Verlag, Berlin.

Trunin, R. F., Podurets, M. A., Simakov, G. V., Popov, L. V. and Moiseev, B. N. (1972). *Sov. Phys.–JETP*, **35**, 550–552.

Tsou, F. K. and Chou, P. C. (1969). *J. Compos. Mater.* **3**, 500–514.

Tsou, F. K. and Chou, P. C. (1970). *J. Compos. Mater.* **4**, 526–537.

Tuler, F. R. and Butcher, B. M. (1968). *Int. J. Fract. Mech.* **4**, 431–437.

Tyunyaev, Yu. N. and Mineev, V. N. (1971). *Sov. Phys.–Solid State*, **13**, 415–417.

Tyunyaev, Yu. N., Mineev, V. N., Ivanov, A. G., Novitskii, E. Z. and Lisitsyn, Yu. V. (1969a). *Sov. Phys.–JETP*, **29**, 98–100.

Tyunyaev, Yu. N., Urusovskaya, A. A., Mineev, V. N. and Ivanov, A. G. (1969b). *Sov. Phys.–Solid State*, **10**, 2683–2688.

Tyunyaev, Yu. N., Mineev, V. N., Ivanov, A. G. and Gutmanas, E. Yu. (1970). *Sov. Phys.–Solid State*, **11**, 2478–2479.

Urlin, V. D. (1966). *Sov. Phys.–JETP*, **22**, 341–346.

Vaidya, S. N. and Kennedy, G. C. (1970). *J. Phys. Chem.–Solids*, **31**, 2329–2345.

Venable, D. and Boyd, T. J. (1965). *In* "Proceedings Fourth Symposium on Detonation", pp. 639–647. Office of Naval Research, ACR-126, Washington, D.C.

von Mises, R. (1958). "Mathematical Theory of Compressible Fluid Flow". Academic Press, New York and London.

von Neumann, J. and Richtmeyer, R. D. (1950). *J. Appl. Phys.* **21**, 232–237.

Wackerle, J. (1962). *J. Appl. Phys.* **33**, 922–937.

Walsh, J. M. and Christian, R. H. (1955). *Phys. Rev.* **97**, 1544–1556.

Walsh, J. M. and Rice, M. H. (1957). *J. Chem. Phys.* **26**, 815–823.

Walsh, J. M., Rice, M. H., McQueen, R. G. and Yarger, F. L. (1957). *Phys. Rev.* **108**, 196–216.

Warnes, R. H. (1967). *J. Appl. Phys.* **38**, 4629–4631.

Warnicka, R. L. (1968). MSC-68-18 (July). Materials and Structures Laboratory, General Motors Corporation.

Warren, B. E. (1959). *In* "Progress in Metal Physics", Vol. 8, pp. 147–202. Pergamon Press, New York.

Warren, B. E. and Averbach, B. L. (1950). *J. Appl. Phys.* **21**, 595–599.

Warren, N. (1973). *J. Geophys. Res.* **78**, 352–362.

Wayne, R. C. (1969). *J. Appl. Phys.* **40**, 15–22.

Weidner, D. J. and Royce, E. B. (1969). *Bull. Amer. Phys. Soc.* **14**, 1170.

Whitworth, R. W. (1965). *Phil. Mag.* **11**, 83–90.

Willardson, R. K., Harmon, T. C. and Beer, A. C. (1954). *Phys. Rev.* **96**, 1512–1518.

Williams, E. O. (1968). *ISA Trans.* **7**, 223–230.

Williams, R. F. and Keough, D. D. (1966). Paper presented at RTI-LRC Thin Film Conference (September 8), Research Triangle Institute, Research Triangle Park, North Carolina.

Williams, R. F. and Keough, D. D. (1967). *Bull Amer. Phys. Soc.* **12**, 1127.

Wong, J. Y. (1969). *J. Appl. Phys.* **40**, 1789–1791.

Wong, J. Y., Linde, R. K. and DeCarli, P. S. (1968). *Nature*, **219**, 713–714.

Wong, J. Y., Linde, R. K. and White, R. M. (1969). *J. Appl. Phys.* **40**, 4137–4145.

Yakushev, V. V. (1972). *Zh. Prikl. Mekh. Tekh. Fiz.* No. 4, 155–161.

Yakushev, V. V., Rozanov, O. K. and Dremin, A. N. (1968). *Sov. Phys.–JETP*, **27**, 213–217.

Yakusheva, O. B., Yakushev, V. V. and Dremin, A. N. (1970). *Sov. Phys.–Dokl.* **14**, 1189–1190.

Young, C., Fowles, R. and Swift, R. P. (1970). *In* Proceedings of the 17th Sagamore Army Research Materials Conference, published 1971: "Shock Waves and the Mechanical Properties of Solids" (Eds J. J. Burke and V. Weiss), pp. 203–223. Syracuse University Press, Syracuse, N.Y.

Young, C., Fowles, R. and Swift, R. P. (1970). Paper presented at the 17th Sagamore Conference on Shock Waves and the Mechanical Properties of Solids (September), Raquette Lake, New York.

Zababakhin, E. I. and Simonenko, V. A. (1967). *Sov. Phys.–JETP*, **25**, 876–877.

Zaitsev, V. M., Pokhil, P. F. and Shvedov, K. K. (1960). *Dokl. Akad. Nauk. SSSR*, **132**, 1339–1340.

Zel'dovich, Ya. B. (1968). *Sov. Phys.–JETP*, **26**, 159–162.

Zel'dovich, Ya. B., Kormer, S. B., Sinitsyn, M. V. and Kuriapin, A. I. (1958). *Sov. Phys.-Dokl.* **3**, 938–939.

Zel'dovich, Ya. B., Kormer, S. B., Sinitsyn, M. V. and Yushko, K. R. (1961). *Sov. Phys.-Dokl.* **6**, 494–496.

Zel'dovich, Ya. B., Kormer, S. B. and Urlin, V. D. (1969). *Sov. Phys.–JETP*, **28**, 855–859.

Zukas, E. G. (1966). *Metals Eng. Quart.* **6**, 1–20.

Note added in proof

Since this review was written, two books dealing with shock-wave phenomena have been published and some new developments on temperature measurement during shock loading are forthcoming. These publications contain many new developments. See "Metallurgical Effects at High Strain Rates" (1973). (Ed. R. W. Rohde, B. M. Butcher, J. R. Holland and C. H. Karnes). Plenum Press, New York and London; "Dynamic Response of Materials to Intensive Impulsive Loading" (1973). (Eds P. C. Chou and A. K. Hopkins). Air Force Materials Laboratory, Wright Patterson AFB, Ohio.

See also several articles to be published in *J. Appl. Phys.* by P. A. Urtiew and R. Grover of Lawrence Livermore Laboratory, Livermore, California, concerning temperature measurement.

X-ray Diffraction Studies up to 300 kbar

W. A. Bassett and T. Takahashi[1]

Departmentment of Geological Sciences and Space Science Center,
University of Rochester, USA

1	Introduction	165
	1.1 Objectives	166
	1.2 X-ray diffraction at high pressures	168
2	Techniques	169
	2.1 Apparatus	169
	2.2 Determination of pressure	178
3	Experimental results	192
	3.1 Metal halides	192
	3.2 Metals and alloys	201
	3.3 Oxides and silicates	218
4	Fixed point calibrations above 100 kbar	239
5	Conclusions	241
	References	243

1 Introduction

Since the first X-ray diffraction study by Jacobs (1938) of metal halides under 3 kbar of helium atmosphere, the working pressure under which substances can be studied by means of X-ray diffraction has increased by as much as two orders of magnitude during the past three decades. At the same time, many different designs of high-pressure apparatus ranging from the simple Bridgman anvils to multi-anvil systems were adapted for X-ray diffraction studies of materials under pressure. As a result of this progress, crystal structures for a number of high-pressure polymorphs have been determined.

[1] Present address: Queens College, The City University of New York, Flushing, N.Y. 11367.

The efforts of our group at the University of Rochester have been directed to the determination of the crystal structures of high-pressure polymorphs, and of the effect of pressure on lattice dimensions (hence the molar volumes) of elements, alloys and compounds with particular emphasis on those pertinent to the geophysical studies of the earth's interior.

1.1 OBJECTIVES

Since the effect of pressure on volume at a constant temperature is directly related to the Gibbs free energy by $(\partial F/\partial P)_T = V$, knowledge of the pressure dependence of volume is of primary importance for the study of thermodynamic stabilities of various phases of substances. In the case of solids, a variety of techniques have been used to determine the pressure–volume relationships: (a) dilatometry (e.g. Bridgman, 1949); (b) shock-wave propagation (e.g. Rice et al., 1958); (c) elastic wave propagation (e.g. McSkimin, 1950); (d) X-ray diffraction (e.g. Jamieson and Lawson, 1962).

The dilatometric techniques permit precise measurements ($\sim \pm 0.1$ per cent in $\Delta V/V_0$) of isothermal P–V relationships for bulk samples up to 45 kbar (Vaidya and Kennedy, 1970) but the pressure measurements become less certain above this pressure. Some of Bridgman's data obtained up to 100 kbar by dilatometry may give pressures as much as 15 kbar too high at 100 kbar.

Shock-wave propagation studies are able to yield pressure–volume data for pressures ranging up to thousands of kilobars. The pressure–volume relationship along a particular P–T path called Hugoniot can be determined from the Rankin–Hugoniot relation (based upon the conservation of energy and momentum) and from an empirical relationship between the propagation velocities of particle and shock waves generated by a detonation of explosives or an expansion of compressed gas. It is commonly assumed that the shock velocity is a linear function of the particle velocity. A phase transformation may be detected as a change in the slope of the particle velocity–shock wave velocity relation. In order to convert the Hugoniot conditions to isothermal conditions, it is necessary to assume the functional form of the Grüneisen parameter, and further, to assume it to be a function of volume only. Therefore, the isothermal P–V relationship derived from the shock-wave experiments may be subject to systematic errors arising from the assumptions used in computational procedures. Such errors become increasingly serious at high pressures and for very compressible materials due to high temperatures of dynamic compression. In spite of such complications, shock-wave experiments are a powerful tool for exploring the $P-V$ relations of various substances at pressures up to several megabars as no other techniques permit measurements at such high pressures.

Measurements of the elastic wave propagation velocities in a bulk sample as a function of pressure and temperature yield highly precise values for the adiabatic bulk modulus and its pressure and temperature derivatives. Such measurements may be made at pressures up to several kilobars and temperatures exceeding 1000°C. The correction for converting the adiabatic to the isothermal bulk modulus is generally less than 5 per cent. An isothermal equation of state of a substance may then be obtained using the values of its volume and bulk modulus at zero pressure and temperature derivatives of the bulk modulus as the boundary conditions. If no phase transformations take place, a reliable $P-V$ relationship may be obtained for pressures up to one or more orders of magnitude greater than those at which the measurements were made.

X-ray diffraction of a polycrystalline specimen while under pressure permits the measurement of lattice parameters as a function of pressure. The isothermal $P-V$ relationship may be obtained from the lattice parameters and from the pressure values determined by some independent means such as the NaCl internal pressure standard. In contrast to the other three methods, which demand rather stringent requirements in quantity and quality of the bulk specimens, the sample required for X-ray diffraction may be in the form of powder, lump or foil and in quantities of milligrams. A phase transformation occurring in the sample may be readily recognized, its crystal structure often identified, and the volume change associated with the phase transformation determined. The maximum working pressure for existing high-pressure X-ray diffraction apparatus is about 300 to 400 kbars at room temperature conditions. While this is approximately one order of magnitude less than the pressure range attainable by the shock-wave method, it exceeds the pressure range for the dilatometric measurements by one order of magnitude and that for the elastic wave propagation measurements by nearly two orders of magnitude. The precision of the measurements attainable by the X-ray diffraction method is comparable to or better than that attainable by the shock-wave methods but is inferior to the precision of the dilatometry and elastic wave propagation measurements. Therefore, the X-ray diffraction method can provide information to bridge the gap between the pressure ranges of the other techniques.

1.2   X-RAY DIFFRACTION AT HIGH PRESSURES

As mentioned earlier, a variety of different methods of applying pressure have been adapted for X-ray diffraction studies of samples under pressure. One of the earliest attempts was that of Cohn (1933) using a Bridgman-type high-pressure container equipped with bakelite and beryllium windows to permit entrance and exit of X-rays. Frevel (1935) applied pressure to a sample in a fluid inside a capillary tube at the centre of a Debye–Scherrer camera. With this arrangement he was able to achieve a pressure of one kilobar. Jacobs (1938) applied pressure to his samples in a special Debye–Scherrer camera by filling the interior of the camera with helium under pressure. With this he not only identified the structure of a high-pressure phase of silver iodide but also measured the continuous change of the molar volume with pressure which he was then able to compare with Bridgman's dilatometric measurements on silver iodide (Bridgman, 1915).

Lawson and Riley (1949) and Lawson and Tang (1950) applied pressures up to 15 kbar to samples in a beryllium cylinder by pumping oil into it. They placed a film around the outside of the cylinder to record the X-rays diffracted by the sample. They also achieved pressures up to 25 kbar by driving two steel pistons together in a hole drilled along the interface between two diamonds.

During the 1950's and 1960's interest in high-pressure X-ray diffraction studies grew rapidly and many devices were developed. These can be divided into six general categories:

1. Piston and cylinder in which the cylinder or a portion of it is constructed of diamond or beryllium (Kasper, 1960; Jamieson, 1961; Bradley et al., 1964; Vereshchagin, 1965).

2. Bridgman anvils arranged so that X-rays enter and leave the sample through the space between the anvils faces. Most instruments of this design use an annular gasket of material transparent to X-rays (Jamieson and Lawson, 1962; Owen et al., 1963; McWhan and Bond, 1964; Perez-Albuerne et al., 1964).

3. A pair of Bridgman anvils, one of which is transparent to X-rays so

that the X-ray beam may be bounced off the sample in a parafocusing geometry (Jamieson, 1961; Davis and Adams, 1962).

4. A tetrahedral press in which gasket material transparent to X-rays is used. The beam may enter through a hole in one of the anvils or it may enter through a gasket between anvils. In either arrangement the diffracted X-rays exit through the gasket between anvils (Barnett and Hall, 1964).

5. Bridgman anvils made of single cyrstal diamonds arranged so that the X-rays enter and leave the sample through the anvils themselves (Piermarini and Weir, 1962; Bassett and Takahashi, 1964; Bassett *et al.*, 1967).

6. High-pressure belt apparatus in which the belt itself has been split along a plane normal to the compression axis and a sheet of beryllium inserted to provide access for X-rays (Freud and Sclar, 1969). The belt apparatus has also been adapted for non-dispersive diffraction (Freud and LaMori, 1969).

In addition to the application of X-ray diffraction to the study of samples under pressure, neutron diffraction has been used with a piston and cylinder device (Brugger *et al.*, 1967) and with a Bridgman anvil device (Brugger *et al.*, 1969).

The highest pressures that have been reported for diffraction studies are 450 kbar, attained by Drickamer *et al.* (1966) for his study of CsCl by means of the supported tungsten carbide anvils, and 330 kbar attained by Bassett *et al.* (1967) by means of the diamond anvils.

2 Techniques

2.1 APPARATUS

2.1.1 *The diamond anvil press*

In the diamond anvil press (Fig. 1) as it has been used at the University of Rochester, a polycrystalline sample is subjected to pressures up to 300 kbar between flat parallel faces of two gem diamonds mounted on the ends of pistons that are driven together by a simple screw. A hole running down the centre of one of these pistons permits the entrance of a finely collimated (50–100 μm) X-ray beam (MoK$_\alpha$), while a slot in the other piston permits X-rays to exit from the sample with a dispersion of $45°$ 2θ each side of the direct beam. A photographic film mounted in a cassette with a radius of curvature of 50 mm records the scattered X-rays.

The anvil faces range from 0.25 to 0.7 mm in diameter depending on the pressures desired. These faces are produced by enlarging the culet faces on 1/8 carat brilliant cut gem diamonds. These diamonds are mounted in recesses in rockers on the ends of the pistons. The rockers, mounted with their axes at right angles to each other, provide

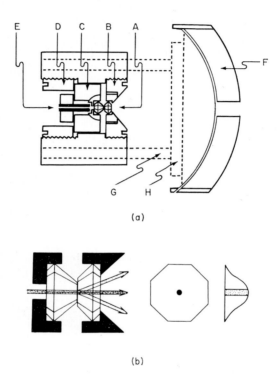

(a)

(b)

Fig. 1. (a) Cross section of the diamond anvil press. A, diamond anvils; B, stationary piston; C, sliding piston; D, driver screw; E, collimator; F, film cassette; G, cassette mounting rods; H, cassette translating bars. (b) Detail of the anvils showing the position of the X-ray beam with respect to the anvils and the pressure distribution at the right.

complete translational and angular freedom for the anvil faces with respect to each other. They permit adjustments so that the faces meet properly and are perfectly parallel to each other.

The X-rays are collimated by a glass capillary which can be moved until the beam is centred in the sample area. The centering is tested by placing photographic film between the diamond anvils and exposing it to the X-ray beam. It is then possible to observe the relationship between the impression by the anvil faces and the spot produced by the X-ray beam (Bassett *et al.*, 1967). The use of glass for the collimator allows light to pass through to the sample so that visual observation of the sample under pressure can be made.

The film cassette is rigidly mounted to the squeezer body by four rods. It is possible to place the sample at the centre of curvature of the cassette by means of gauges. A platelet of polycrystalline NaCl placed on the outside of the diamond anvil facing the film serves to mark the film with its diffraction pattern. This diffraction pattern is then used as a basis for correcting errors resulting from changes in camera and film dimensions. The sample-to-film distance is calculated from the diffraction pattern of a known sample, thereby eliminating most of the other sources of systematic error.

Measurements have been made with both filtered molybdenum radiation and monochromatized molybdenum $K_{\alpha 1}$ radiation. The filtered radiation is satisfactory for samples with high atomic number or samples that are being studied at low pressure. The monochromatic radiation eliminates spots due to Laue reflections from the single crystal diamond anvils and greatly reduces the background darkening that results from the scatter of the continuum part of the X-ray spectrum by the anvils.

Heating of samples up to temperatures as high as 310°C while under pressure has been achieved by means of a ceramic heater placed immediately around the pistons (Bassett and Takahashi, 1965). A lever device is used in this arrangement so that the load can be maintained constant by an external spring that is kept cool during heating. Temperature of the sample is measured by means of a thermocouple placed in contact with the diamond anvils.

2.1.2 *Experimental accuracy*

There are five sources of error to contend with in the diamond anvil press: (1) reading error; (2) change in film dimension; (3) change in sample-to-film distance; (4) pressure gradient effects; (5) change in wavelength due to change in geometry of the monochromator. Repeated readings on the same pattern show a standard deviation of 0.03 per cent for *d*-spacing. Errors in film dimension and sample-to-film distance are corrected by means of the film marking produced by X-ray diffraction of an NaCl platelet fastened to the outside of the anvil facing the film. After this correction, they introduce a random error that is also approximately 0.03 per cent of the *d*-spacing. The sample-to-film distance is calculated from the diffraction pattern of a known material placed between the anvils at 1 bar pressure or it is calculated directly from the patterns produced by the external NaCl and the thickness of the anvil on the film side of the sample. The same errors are involved in the sample-to-film distance determination as in the high-pressure lattice parameter determinations and are thus cancelled out. The monitoring of the instrument geometry by means of the external NaCl sometimes indicates a shortening of the sample-to-film distance during a series of runs. This is almost certainly due to the seating of the diamond anvil into the soft stainless steel on the film side of the sample and must be corrected.

Since the external NaCl is able to resolve $K_{\alpha 1}$ and $K_{\alpha 2}$, it provides a means for determining the spectrum of the radiation used. All photographs produced with a monochromatized X-ray beam show that the radiation was pure $MoK_{\alpha 1}$. Therefore, variation of wavelength can be eliminated as a source of error. Absorption is not considered a serious source of error because of the very thin tabular shape of the sample and because the sample-to-film distance is determined by means of an X-ray diffraction pattern of known sample and is therefore subject to the same error. Line broadening indicates that the range of pressure within the X-ray beam is ordinarily not more than 10 per cent of the central pressure. In some instances of excessive line breadth, lattice parameter determinations have been based on measurements made at the outer edges of the diffraction lines. This guarantees that the measurements are based on X-ray scattered by the portion of the sample at the highest pressure. This portion of the sample is not only subjected to the highest pressure and lowest pressure gradient, but also is least affected by the orientation effects of the shearing that takes place as sample is extruded from between the anvil faces. This procedure may be necessary if the shearing has a different effect on the orientation of the sample from that of the standard.

When all of the possible corrections are made, a random error of approximately 0.1 per cent remains. This results in a volume error of 0.3 per cent. This in turn leads to a precision of approximately 2.5 per cent in pressure measurements when NaCl is used as the pressure standard.

2.1.3 *Deformation of anvil faces*

According to our observations, the deformation of the diamond anvils is elastic and therefore exists only while they are under load. The only permanent changes which take place in the diamond anvils are fractures. The causes of fracturing are mainly overloading and mis-alignment. There is no evidence at present that frequency of loading has any effect on fracturing of diamond.

Plastic deformation of the soft stainless steel anvil seat does take place. Therefore, occasional realignment of the anvils as well as corrections for changes in sample-to-film distance are required for accurate measurements of the lattice parameters.

There are four sources of information on the elastic deformation of the diamond anvils: (1) birefringence of the diamonds under load; (2) interference between monochromatic light rays reflected from the two anvil faces; (3) elongated Laue spots produced by the anvils under load; (4) the relationship between the measured and the applied pressure, force/area.

The birefringence of the diamonds can be observed between crossed polars in a polarizing microscope when an isotropic substance is placed between the diamonds and load is applied. Since the observation is made looking along the axis of compression, the birefringence appears as a dark cross against a light background (for example, see Fig. 4 in Bassett and Takahashi, 1965). As load is increased, the background becomes coloured and the shape of the cross becomes distorted. When the load is released, the birefringence disappears completely and the diamonds resume their optical isotropy.

Interferometry can be used to observe deformation of the anvils by means of observations of interference bands produced when monochromatic light passing down through the upper diamond is reflected from the two anvil faces. If a chip of metal is placed at the centre of the sample area and the diamonds are driven together, interference bands appear at the edge of the sample area and proceed inward indicating that the anvil faces are cupping around the chip. When the load is released, the bands disappear.

When a diffraction pattern is obtained by passing an unmonochromatized X-ray beam *perpendicular* to the axis of compression, Laue spots are produced by the portions of the diamond anvils immediately adjacent to the sample and therefore subject to the maximum deformation.

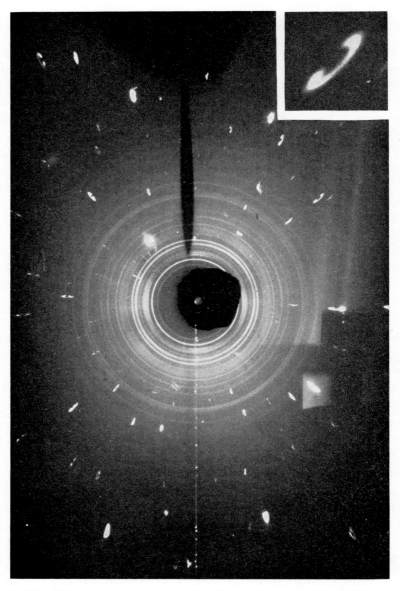

Fig. 2. This diffraction pattern was produced by passing an unfiltered X-ray beam (Mo target) perpendicular to the axis of compression in a diamond anvil press. The rings were produced by the polycrystalline sample and the spots were produced by the portions of the diamond anvils adjacent to the sample. The insert is an enlargement of one of the Laue spots from the diamond anvils showing that the spots are "c" shaped. The "c" shape of the Laue spots is attributed to the elastically strained diamond crystal.

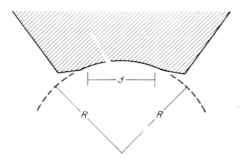

Fig. 3. The proposed distortion of the anvil faces producing the Laue spots in Fig. 2.

The pattern shown in Fig. 2 was produced in this geometry with a sample under a pressure of approximately 100 kbar. The "c"-shaped Laue spots indicate warped anvil surfaces having the configuration shown in Fig. 3. The following equations can be used to calculate the radius of curvature, R, of the warped diamond surface:

$$R = 360 d / 2 \pi \Delta \theta$$
$$\Delta \theta = \tfrac{1}{2}[\tan^{-1}(S_2/G) - \tan^{-1}(S_1/G)]$$

where S_1 is the distance from the centre of the film to the nearest part of the "c", S_2 is the distance from the centre of the film to the farthest part of the "c", G is the sample-to-film distance, $\Delta \theta$ is the maximum change in the orientation of the lattice along the anvil face, d is the distance between the points of greatest difference of orientation along the anvil face and is assumed to be half the diameter of the anvil face as shown in Fig. 3. The following set of values pertain to the photograph shown in Fig. 2: $S_1 = 41$ mm, $S_2 = 45$ mm, $G = 50$ mm, and $d = 0.25$ mm. The radius of curvature, R, calculated from these is 12 mm.

Finally, information concerning the deformation of the diamond anvils can be derived from the relationship between the applied pressure (force/area) and the pressure measured at the centre of the sample area by means of an internal standard. Figure 4 shows the relationship between the measured pressure and the applied pressure for a mixture of the spinel phase of Fe_2SiO_4 and NaCl. Initially the central pressure is approximately twice the applied pressure. As load is increased, the ratio of central pressure to applied pressure decreases. This indicates that the pressure distribution becomes flatter, and, as a consequence, the pressure at the centre of the anvil area approaches the applied pressure.

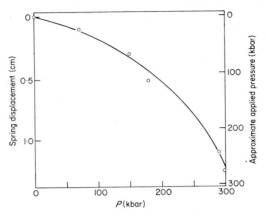

Fig. 4. Sample pressure determined by NaCl standard as a function of load and applied pressure (force/area) in a diamond anvil cell having anvil faces 0.32 mm in diameter.

All of the evidence presented here indicates that the diamond anvils deform elastically under load so that their faces become cup shaped around a lenticular sample. One of the most beneficial effects of this anvil deformation is the diminishing of the pressure gradient in the central portion of the sample area. This is particularly noticeable in the manner in which the high-pressure phase of NaCl appears at 300 kbar. It appears simultaneously over nearly one-half of the sample area in the centre. This behaviour is in striking contrast with that of phase transformations taking place at lower pressures. In these cases the high-pressure phase makes its first appearance as a small spot at the centre of the sample area and increases in size as the load is increased (Bassett and Takahashi, 1965). The flat pressure distribution at 300 kbar permits the use of a rather large collimator without line broadening due to pressure gradient.

2.1.4 *Comparison of the diamond anvil press with other high-pressure X-ray diffraction instruments*

Each high-pressure X-ray diffraction instrument has advantages and disadvantages unique to it. The diamond anvil press has the following advantages:

1. It is able to achieve pressures up to 300 kbar, and can be operated continuously for as long as several months in a pressure range that

lies beyond the capabilities of most high-pressure X-ray diffraction instruments.

2. Samples as small as a few milligrams can be used.
3. Visual observation can be made on the sample while it is under pressure.
4. Very little distortion of the geometry of the instrument takes place during a series of measurements. The error which results from the small amount of distortion which does occur is easily corrected by means of the external NaCl described in the previous section.
5. The design of the instrument is simple. There is less chance, therefore, of unexpected and undetected sources of error.

The following are among some of the disadvantages of the diamond anvil press:

1. The small size of the sample requires that exposure times be rather long. They range from 150 h for low-pressure determinations to 400 h for determinations near the upper limit of the pressure range. The longest exposure time used was 1000 h.
2. The extrusion of the sample from between the anvils results in some preferred orientation of the sample. The amount of orientation differs from material to material. Thus, for some samples intensity measurements are unreliable and caution must be used in making structure determinations. When the orientation effect in the sample and in the standard are very different and when there is a fairly large pressure range within the X-ray beam, skewing of the intensity profiles for each line may introduce an error in lattice parameter measurements. This is easily detected and corrected by basing 2θ measurements on the outermost edges of the lines. The outer edges are produced by the portion of the sample and standard which are at highest pressure at the centre of the sample area and are therefore subjected to the least amount of shear stress due to extrusion of the sample from between the anvils.

Some of the compression data for haematite and stishovite were measured with the tetrahedral press at Brigham Young University. These are compared with data obtained with the diamond press in Figs 29 and 36. In general we feel that the agreement is good though the scatter is a little greater in the tetrahedral press data. Also, there are five determinations of stishovite which significantly disagree with the data from the diamond press. We attribute this disagreement to a deformation of the tetrahedral press when the sample gaskets become essentially incompressible. Such a deformation may introduce a significant error in measurements of so incompressible a material as stishovite but not in more compressible samples such as NaCl. The instrument at Brigham Young University has since been modified to prevent such a deformation.

2.2   DETERMINATION OF PRESSURE

Since the pressure at the centre of the sample area in the diamond anvil press is not equal to the applied pressure or some predictable function of the applied pressure, it is necessary to use an internal standard for determining the pressure. This is accomplished by making an intimate mixture of the sample and the standard both in powder form. Such a mixture produces an X-ray diffraction pattern exhibiting the diffraction lines for both substances. The lattice parameters of the standard substance compressed along with the sample can be determined with an accuracy of ± 0.1 per cent. If the standard substance has a small yielding strength against shearing, it should flow plastically under small differential stress, and hence the pressure to which the standard substance is subjected should be equal to the pressure to which the sample is subjected. Therefore, the substance suitable for the internal pressure standard should have not only a well-established P–V relationship, but also a small yielding strength against shear. Although a number of soft metals are easily deformable and their P–V relationships are well known, they are opaque and thus prevent the optical observations of the sample in transmitted light. Furthermore, some soft and deformable metals such as alkali metals are chemically active and may react with the sample or with the atmosphere to produce compounds whose diffraction patterns interfere with the diffraction patterns of the phases being studied. Therefore, two simple ionic solids, NaCl and CsI, were chosen as suitable standards for our investigations. These two substances have the following advantages as pressure standards:

1. The small bulk moduli of NaCl and CsI (237.4 and 118.9 kbar respectively at 1 bar and 25°C) provide good sensitivity since a small change in pressure results in a relatively large change in volume.
2. The cubic structures of NaCl (B1) and CsI (B2) yield X-ray diffraction patterns with relatively few lines, thus minimizing the chance of

interference with the sample pattern. In addition, each reflection in the pattern yields an independent value for the unit cell dimension because of the high symmetry.

3. The low shear strength of each of these salts produces a more hydrostatic pressure environment than solids having higher shear strengths.
4. Both salts are transparent to visible light.
5. Both salts have been the subject of a great number of theoretical and experimental studies, so that most of the properties necessary for calculation of the $P-V$ relation are available from the literature.
6. Both salts are chemically stable under normal laboratory conditions.

2.2.1   The NaCl pressure scale

The determination of pressure in all the compression measurements carried out at our laboratory is based upon the $P-V$ relation for NaCl calculated by Weaver et al. (1971) using the vibrational formulation of the Hildebrand equation (Table 1). The results of several experimental

TABLE 1

The calculated $P-V$ relation for NaCl at 25°C by means of the vibrational Hildebrand equation (Weaver et al., 1971)

V/V_0	P(kbar)	V/V_0	P(kbar)
1.00	0.00	0.720	172.0
0.950	13.9	0.710	185.3
0.925	22.5	0.700	199.5
0.900	32.5	0.695	206.9
0.880	41.6	0.690	214.6
0.860	51.8	0.685	222.6
0.840	63.4	0.680	230.8
0.820	76.4	0.675	239.3
0.800	91.1	0.670	248.1
0.790	99.1	0.665	257.2
0.780	107.7	0.660	266.6
0.770	116.8	0.655	276.3
0.760	126.5	0.650	286.3
0.750	136.8	0.645	296.7
0.740	147.8	0.643	301.0
0.730	159.5		

The B1 phase transforms to the B2 phase at $V/V_0 = 0.643$ (Bassett et al., 1968).

and theoretical investigations of the *P–V* relation for NaCl are in the literature. The *P–V* relation used in our studies will be compared with those, and the reasons for favouring this pressure scale will be explained.

The *P–V* relation for NaCl was repeatedly investigated by Bridgman (1940, 1945 and 1948) up to 100 kbar at room temperature by means of his straight piston-cylinder dilatometric apparatus. His data are compared in Fig. 5 with the *P–V* relation calculated by Weaver *et al.* (1971). It is seen that Bridgman's 100-kbar and 40-kbar data agree with

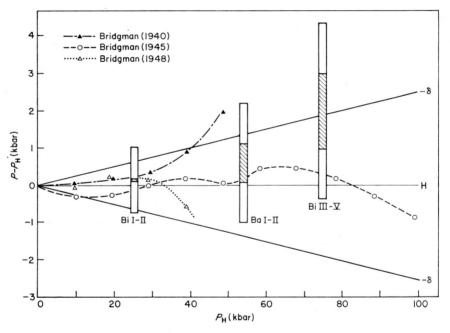

Fig. 5. A comparison between the vibrational formulation of the Hildebrand equation (H) (Weaver *et al.*, 1971) and other compression data for NaCl. The solid lines marked with $+\delta$ and $-\delta$ indicate the limit of the computational error caused by the uncertainties in the values used for the boundary conditions. The broken lines are the results of three compression studies by Bridgman. The vertical bars indicate the difference between the pressures for the Bi I-II, Ba I-II and Bi III-V transformations obtained from the data of Jeffrey *et al.* (1966) using the *P–V* relation of Weaver *et al.* (1971) for NaCl, and the pressures of these transformations determined by Kennedy and LaMori (1961), Haygarth *et al.* (1967), and Haygarth *et al.* (1969) using a straight piston-cylinder device. The experimental uncertainties for Jeffrey *et al.* are indicated by the clear bars, and those for the determinations of the transitions are indicated by the shaded bars.

the values of Weaver *et al.* within ± 1 kbar. On the other hand, Bridgman's 50-kbar data deviate beyond the uncertainty of the calculated *P–V* relation at pressures greater than 40 kbar. With the exception of this pressure range, our pressure scale is consistent with Bridgman's data within 1 kbar.

The Hugoniot equation of state for NaCl has been obtained up to several hundred kilobars by means of the shock-wave technique by Christian (1957), Fritz *et al.* (1971) and others. Although a linear relationship between the shock and particle velocities is commonly observed for a number of substances, the experimental data for NaCl do not show an unequivocal linear relationship for the shock and particle velocities (Fritz *et al.*, 1971). Since the isothermal *P–V* relation reduced from the shock data depends on the form of the shock velocity–particle velocity relation as well as the form of the volume dependence of the Grüneisen parameter, the isothermal *P–V* relation reduced from the shock data suffers from the uncertainties resulting from the basic assumptions used in the data reduction. The results of Fritz *et al.* (1971) are compared with the *P–V* relation of Weaver *et al.* (1971) in Fig. 6. The values resulting from the linear and quadratic fit of the shock and particle velocities data do not differ from the results of Weaver *et al.* by more than 5 kbar, and the differences are mostly within the uncertainty of the calculation indicated by $\pm \delta$ in Fig. 6. Weaver (1971a) critically compared the *P–V* relation obtained by Weaver *et al.* (1971) with the shock compression data of Christian (1957), Alt'shuler *et al.* (1961), Hauver and Melani (1970), and Fritz *et al.* (1971). He found that there is no compelling reason to choose one over the others.

Decker (1965, 1966) calculated the equation of state for NaCl using the vibrational Mie–Grüneisen equation. Perez-Albuerne and Drickamer (1965) calculated the equation of state for NaCl using the thermal Hildebrand equation. For these studies, the parameters determined at or near 25°C and 1 bar were used as the boundary conditions. The discrepancy between the results of Decker (1966) and Weaver *et al.* (1971) is due to the difference in the values for the initial bulk modulus used for a boundary condition: an older value of 234.2 ± 1.8 kbar at 25°C (Bartels and Schuele, 1965) was used by Decker, whereas a newer value, 237.4 ± 1.7 kbar, which is the mean of the determinations by Slagle and McKinstry (1967) and by Haussühl (1960), was used by Weaver *et al.* (1971). The revised *P–V* relation calculated by Decker *et al.* (in press) using the new value for the bulk modulus is virtually

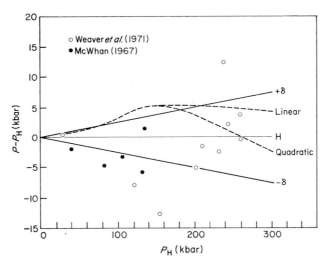

Fig. 6. A comparison between the vibrational formulation of the Hildebrand equation (H) (Weaver et al., 1971) for NaCl and other P–V data. The dashed lines are linear and quadratic fits to the shock velocity–particle velocity data (Fritz et al., 1971). The circles are static compression data for NaCl calculated from volume comparisons of NaCl and MgO using the calculated P–V relation for MgO with the Murnaghan equation (see text for the parameters used).

in agreement with those of Weaver et al.. The discrepancy between the results of Perez-Albuerne and Drickamer (1965) and those of Weaver et al. (1971) is mainly due to the difference in the forms of the Hildebrand equation used for their calculations. The former investigators used the thermal formulation, whereas the latter investigators used the vibrational formulation. According to Fumi and Tosi (1962), the vibrational Hildebrand equation is more appropriate for temperatures near the Debye temperature. Since the Debye temperature for NaCl, 290 ± 12 K, is close to 25°C, the vibrational formulation is preferred for the computation of the P–V relation at 25°C.

Thomsen (1970) used the Lagrangian formulation of strain for relating the pressure to the volume of solid, instead of using the Eulerian formulation which has been used for the derivation of the widely accepted Birch–Murnaghan equation of state. In the Lagrangian formulation the strain generated in a solid as a result of compression is described with reference to the initial or undeformed coordinates, whereas in the Eulerian formulation, the strain is described with reference to the deformed coordinates. Following the treatments of anharmonicity in solids by Leibfried and Ludwig (1961), Thomsen derived a self-consistent equation which includes the contributions of anharmonic lattice vibrations. Since the value of $(\partial K^2/\partial P^2)_T$, where K is the bulk modulus of solid, is not available for NaCl, he used the shock Hugoniot data of Fritz et al. (1971) to obtain this value, and then calculated the P–V relation for NaCl at 25°C. His results are in satisfactory agreement with those of Weaver et al. (1971) as listed in Table 2.

TABLE 2

Direct comparison of the effect of pressure on the volume of MgO and NaCl

	MgO				NaCl	
V/V_0	Murnaghan equation[a] kbar	Birch–Murnaghan equation[b] kbar	Thomsen equation[c] kbar	V/V_0	Hildebrand equation[d] kbar	Thomsen equation[e] kbar
0.983	28.5 ± 5	28.5 ± 5	28 ± 4	0.911	28 ± 1	28
0.940	114 ± 6	114 ± 6	113 ± 6	0.764	122 ± 2	124
0.928	142 ± 7	142 ± 7	140 ± 7	0.735	154 ± 3	154
0.907	196 ± 7	195 ± 7	191 ± 8	0.700	199 ± 3	198
0.902	210 ± 8	208 ± 8	203 ± 8	0.692	211 ± 3	208
0.895	230 ± 8	228 ± 8	225 ± 8	0.679	232 ± 3	228
0.888	251 ± 8	249 ± 8	246 ± 8	0.671	246 ± 4	240
0.887	254 ± 8	251 ± 8	—	0.676	238 ± 4	—
0.884	265 ± 9	262 ± 9	—	0.662	262 ± 4	254
0.883	267 ± 9	264 ± 9	260 ± 9	0.664	259 ± 4	—

[a] $P = (K_0/K_0')\,[(V/V_0)^{-K_0'} - 1]$, $K_0 = 1599$ kbar, $K_0' = 4.52$; Anderson and Andreatch (1966).
[b] $P = 1.5\,K_0\,[(V/V_0)^{-7/3} - (V/V_0)^{-5/3}]\{1 - \xi[(V/V_0)^{-2/3} - 1]\}$, $K_0 = 1599$ kbar and $\xi = -0.39$.
[c] Thomsen (1970).
[d] Weaver et al. (1971).

In order to test the reliability of those P–V relations for NaCl described above, the effect of pressure on the lattice parameter of NaCl was directly compared with that of MgO at various pressures by means of X-ray diffraction of an intimate mechanical mixture of MgO and NaCl compressed in a diamond-anvil cell. MgO was selected for the following reasons: (1) highly refined values for its bulk modulus and pressure derivative of bulk modulus are available; (2) it is available in a pure form; and (3) its P–V relation can be calculated with a high precision using simple equations. Since MgO compresses only to 0.87 in V/V_0

at 300 kbar, the Murnaghan or the Birch–Murnaghan equations should yield a reliable P–V relation in this compression range. The results of the simultaneous compression of MgO and NaCl are listed in Table 2. The MgO pressure values shown in the Table were calculated by means of the Murnaghan equation, the Birch–Murnaghan equation and the fourth-order anharmonic equation of Thomsen (1970), and the NaCl pressure values were calculated from the vibrational Hildebrand equation by Weaver et al. (1971). The MgO pressures calculated from the Murnaghan equation and the NaCl pressures were also plotted in Fig. 6 along with the data obtained by McWhan (1967) up to 120 kbar. The pressure values obtained for MgO and NaCl are in agreement within a root mean square value of ± 7 kbar, and do not show an apparent systematic difference between them. Therefore, the P–V relation for NaCl calculated by Weaver et al. (1971) is internally consistent with the P–V relation for MgO. However, the precision of the data is not good enough to provide a criterion for discriminating among the NaCl P–V relations calculated by Decker et al. (in press), Perez–Albuerne and Drickamer (1965), Thomsen (1970), Fritz et al. (1971), Weaver et al. (1971).

Additional support for the NaCl pressure scale by Weaver et al. has been obtained from the agreement between the ultrasonically determined bulk modulus values and the bulk modulus values calculated from the P–V data of various samples using this NaCl pressure scale. For example, the isothermal bulk modulus of an almandite garnet determined by Soga (1967) by means of the ultrasonic method is 1.757 Mbar, and is in good agreement with the value of 1.74 ± 0.07 Mbar determined by Takahashi and Liu (1970) using X-ray diffraction and the NaCl pressure scale of Weaver et al. (1971). Similarly, the isothermal bulk modulus of elemental rhenium determined by Shepard and Smith (1965) by means of the ultrasonic technique is 3.587 Mbar at room temperature, and is in good agreement with the value of 3.5 ± 0.15 Mbar determined by Liu et al. (1970) using the X-ray technique and the NaCl pressure scale of Weaver et al. (1971).

2.2.2 The CsI pressure scale

Although NaCl has proved to be a very satisfactory pressure standard, CsI has some features which make it attractive also as a pressure standard, especially under certain circumstances: (1) its mean atomic number is 54, approximately four times that of NaCl—therefore, it has a scattering power nearly four times that of NaCl; (2) its zero-pressure bulk modulus, 118.9 kbar at 25°C, is about half that of NaCl and the CsI pressure scale is therefore nearly twice as sensitive as the NaCl pressure scale; (3) it has a simple B2 (CsCl type) crystal structure which does not undergo a phase transformation within the pressure–temperature range of our present capabilities; (4) highly refined elastic data are available at low pressures. Among its disadvantages are its high mass absorption coefficient and its deliquescence. In the diamond press the sample thickness is small enough for the high mass absorption coefficient not to present a serious problem. The deliquescence requires that special arrangements be made to assure a dry atmosphere for the sample.

Bridgman (1940, 1945) studied the effect of pressure on the volume of CsI at room temperature using two different modifications of his apparatus. Although his earlier study up to 50 kbar yielded somewhat smaller compression values (4 per cent in $\Delta V/V_0$) than his later study up to 100 kbar, these two sets of data are in agreement within 0.5 per cent or better in V/V_0. Barsch and Chang (1971) calculated the P–V relation for CsI up to 500 kbar by means of the Murnaghan and the Birch–Murnaghan equations using 118.9 ± 0.5 kbar for zero-pressure bulk modulus, K_0, 5.93 ± 0.08 for $(\partial K/\partial P)_\mathrm{T}|_{p=0}$, and -0.073 ± 0.008 kbar^{-1} for $(\partial^2 K/\partial P^2)_\mathrm{T}|_{p=0}$.

Their calculated values are essentially in agreement with Bridgman's data up to 50 kbar, but deviate to higher pressures as much as 20 per cent in the pressure range of 50 to 100 kbar. On the other hand, their calculated values coincide with the shock Hugoniot data of Christian (1957). Since the Hugoniot data include the effect of thermal expansion of CsI samples shocked to high temperatures, the Hugoniot data at a given pressure should give a greater volume than that for the 25°C isotherm at the same pressure. Therefore, it appears that Barsch and Chang (1971) had overestimated the volume at the 25°C isothermal conditions.

Since the P–V relation for CsI had not been well established above 100 kilobars, the measurement of the effect of pressure on the volume of CsI was undertaken in our laboratory by Hammond (1969) using NaCl as the pressure standard. The experimental data are shown in Fig. 7 and are compared with the isothermal data obtained by Bridgman (1940, 1945) at room temperature, with the shock Hugoniot data of Christian (1957), and with the results of the calculation by Barsch and Chang (1971). In an effort to explain his experimental data, Hammond (1969) assumed that the cohesive lattice energy u is in the form of

$$u = -A/r - C/r^6 - D/r^8 + Be^{-r/q},$$

and applied it to the thermal Hildebrand equation. The definitions and the numerical values for the parameters in the equation above are

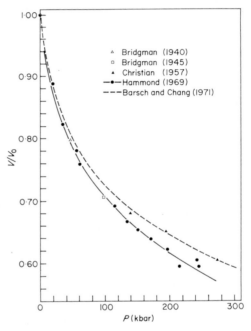

Fig. 7. A comparison of compression data for CsI from five sources.

listed in Table 3. For the boundary condition, the K_0 value of 119.1 kbar (at 23°C) obtained by Chang and Barsch (1967) was used. The $P-V$ relation thus calculated yields a K_0'' value of 5.84, which is somewhat smaller than the experimental value of $5.93 + 0.08$ obtained by Barsch and Chang (1971). However, Hammond found that his calculations yielded about 13 per cent greater pressure values (Model A in Fig. 8) than the experimental data for compressions up to $V/V_0 = 0.60$, when Mayer's (1933) value for the dipole–dipole van der Waal's constant, C, of 2970×10^{-60} erg cm^6 was used. When Hajj's (1966)

TABLE 3

Constants used in calculation of the $P-V$ relation for CsI using the Hildebrand equation

Parameter	Values	Uncertainty %	References
α_r, Madelung constant	1.76267		Tosi (1964)
e, electron charge	4.803×10^{-10} esu		
r, nearest neighbour distance	3.9557 Å		Swanson et al. (1953)
C, dipole–dipole van der Waals constant	2970×10^{-60} erg cm^6	± 20	Mayer (1933)
	4915×10^{-60} erg cm^6	± 20	Hajj (1966)
D, dipole–quadrapole van der Waals constant	5800×10^{-76} erg cm^8	± 100	Mayer (1933)
α, volume thermal expansion coefficient	1.212×10^{-4} °C^{-1}	± 6	Johnson et al. (1955)

value of 4915×10^{-60} erg cm^6 for the van der Waals constant, C, was used, the difference between the calculated values and the experimental values was reduced to about 7 per cent (Model A$'$ in Fig. 8). However, the value for K_0' decreased to 5.61 when Hajj's constant was used. Therefore, the simple Born–Mayer type interatomic potential which includes two adjustable parameters explains the experimental data to no better than 5 per cent in pressure.

Figure 9 shows a comparison between the experimental results and the various sets of calculated values. The curves M1 and M2 were calculated by Barsch and Chang (1971) using the first- and second-order Murnaghan equations respectively, and the B1 and B2 curves were calculated using the first- and second-order Birch–Murnaghan equations (Barsch and Chang, 1971). The curve B2$'$ indicates the $P-V$ relation calculated from the second-order Birch–Murnaghan equation using a K_0'' value of -0.10 kbar^{-1} instead of the experimental value of -0.073 kbar^{-1} obtained by Barsch and Chang (1971).

Following Barsch and Chang (1971), Hammond (1969) used the Birch–Murnaghan equation which is based on the Eulerian scheme of the description for strain (Murnaghan, 1937; Birch, 1947), and fitted it to his experimental data to obtain the value for K_0'' assuming the K_0 and K_0' values of Barsch and Chang (1971). The K_0'' value thus obtained is -0.13 kbar^{-1}, which is nearly twice as great (in its absolute value) as the experimental value and lies outside of the experimental uncertainty of Chang and Barsch (1967). Therefore, the compression data of Hammond (1969) cannot be reconciled with the sonic data of Barsch and Chang (1971) when the Birch–Murnaghan equation is used.

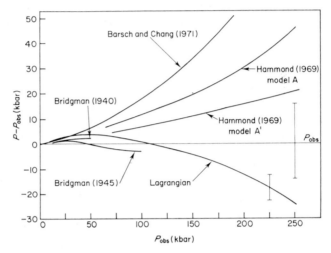

Fig. 8. A comparison between Hammond's compression data for CsI, the determinations of Bridgman (1940 and 1945) and the calculated P–V relations. The P–V relation calculated by Barsch and Chang (1971) is based on the second-order Birch–Murnaghan equation; the two curves by Hammond (1969) are based on the Hildebrand equation with the Born–Mayer lattice energy; and the one marked Lagrangian is based on the Lagrangian equation using the ultrasonic data of Barsch and Chang (1971).

As an alternative to the Eulerian scheme, Murnaghan (1937) and Birch (1947) used the Lagrangian scheme for describing the strains produced in solids by hydrostatic compression. However, since their pioneering work, the equation of state based on the Lagrangian scheme did not gain popularity as much as the Eulerian-based Birch–Murnaghan equation. The lesser popularity of the Lagrangian formulation is probably due to the fact that it converges at a slower rate than the Eulerian formulation, and hence that the values for the pressure derivatives of higher orders are needed to obtain reliable results. The significances of the Langrangian formulation are fully discussed by Thomsen (1970).

The equation of state based on the Lagrangian formulation of strain

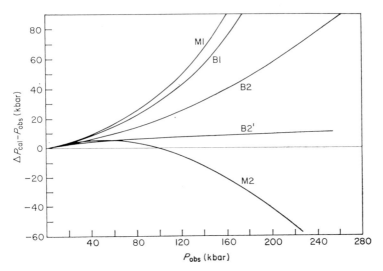

Fig. 9. Comparison between the experimental values of Hammond for CsI at room temperature, and the P–V relations calculated with the first (M1) and second (M2) order Murnaghan equations and the first (B1) and second (B2) order Birch–Murnaghan (Eulerian strain) equations using $K_0 = 118.9$ kbar, $K_0' = 5.93$, and $K_0'' = -0.073$ kbar^{-1} (Barsch and Chang, 1971). The curve B2′ was calculated using the second-order Birch–Murnaghan equation with a K_0'' value of -0.10 kbar^{-1}.

can be derived readily from the following relationship according to the finite strain theory:

$$\eta = \tfrac{1}{2}[(V/V_0)^{2/3} - 1]$$

$$F = \sum_{i=1}^{n} A_i \eta^i,$$

and

$$P = (\partial F/\partial V)_T,$$

where η is strain, F is free energy, and A_i is a coefficient to be evaluated. When n is set to 4, we obtain

$$P = -\tfrac{3}{2} K_0 (V/V_0)^{-1/3} [(V/V_0)^{2/3} - 1] \{1 - \tfrac{3}{4} K_0'[(V/V_0)^{2/3} - 1]$$
$$+ \tfrac{1}{24}[(V/V_0)^{2/3} - 1]^2 [9(K_0' + K_0 K_0'' + K_0'^2) - 1]\},$$

where K denotes the isothermal bulk modulus, the primes denote pressure derivatives at constant temperature, and the subscript 0 denotes the values evaluated at zero pressure. The P–V relation calculated for CsI with this equation using the values of $K_0 = 118.9$ kbar,

$K_0' = 5.93$ and $K_0'' = -0.073$ kbar^{-1} is shown in Fig. 8 and the values are listed in Table 4. Although the Lagrangian strain equation exhibits a fair agreement with the experimental data up to 150 kbar, it deviates

TABLE 4

The P–V relation for CsI determined by Hammond (1969) and those calculated by means of the equation of state based on the Lagrangian scheme for strain and the Hildebrand equation

	Pressure (kbar)			
V/V_0	Lagrangian[a]	Hildebrand[b]	Experimental Hammond (1969)	Fritz–Thurston[c]
1.00	0.0	0.0	0	0.0
0.95	7.1	7.1	7	7.2
0.90	17.0	16.8	14	16.8
0.85	30.7	30.7	28	30.8
0.80	49.0	49.3	46	49.5
0.75	73.4	75.9	70	75.2
0.70	105.6	113.4	105	109.0
0.65	148.0	167.1	154	154.0
0.60	201.0	244.9	218	212.3
0.55	264.4	—	—	287.7

[a] $K_0 = 118.9$ kbar (25°C), $K_0' = 5.93$ and $K_0'' = -0.073$ kbar^{-1}: Barsch and Chang (1971).

[b] Hammond (1969). Hajj's (1966) value of 4915×10^{-60} erg cm^6 was used for the dipole–dipole van der Waals constant C.

[c] Fritz–Thurston equation using the elastic parameter values of Barsch and Chang (1971).

widely at pressures exceeding 150 kbar to a negative direction. At 250 kbar the deviation is about 10 per cent in pressure, and is similar to that for Model A′ in magnitude but is in the opposite direction. Therefore, the Langrangian equation combined with the elastic parameters of Barsch and Chang (1971) fails to explain the experimental data at pressures greater than 150 kbar, which is equivalent to about 1.5 times the bulk modulus, 118.9 kbar.

   In Fig. 10 the experimental data of Hammond (1969) are compared with the compression data of Bridgman (1940), the shock compression data of Christian (1957), the recent static measurements by Vaidya and Kennedy (1971), and the values calculated from the semi-empirical

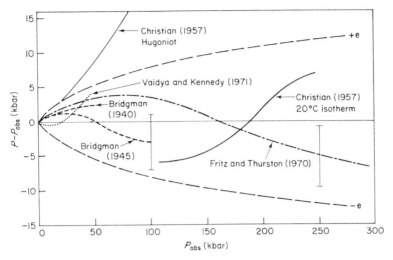

Fig. 10. Comparison of the experimental values for CsI of Hammond (1969) and of Bridgman (1940, 1945) with the shock Hugoniot of Christian, the 20°C isotherm reduced from the Hugoniot and the Fritz–Thurston equation trajectory. The Fritz–Thurston equation trajectory was calculated using the ultrasonic elastic parameter values of Barsch and Chang (1971). The error bar indicates the uncertainty resulting from the experimental uncertainties in the elastic parameter values. The 20°C isotherm of Christian was reduced from his Hugoniot data assuming a constant Grüneisen parameter. The uncertainty of the experimental data of Hammond (1969) is indicated by the dashed curves marked with + e and − e.

equation derived by Fritz and Thurston (1970) using the ultrasonic elastic parameters, K_0, K_0', and K_0'' obtained by Barsch and Chang (1971):

$$V/V_0 = \left\{ a[mp^2 + (1+A+am)p + a]^{-1} \right.$$
$$\left. \times \left[\frac{4am + 2mp[q^{1/2} + (1+A+am)]}{4am - 2mp[q^{1/2} - (1+A+am)]} \right]^{(1+A-am)/q^{1/2}} \right\}^{1/2m}$$

$$m = \lim_{p \to \infty} (\mathrm{d}K/\mathrm{d}P) = 5/3 \text{ for nonrelativistic Fermi gas}$$

$$a = -2(K_0' - m)/K_0 K_0'', \qquad A = a(K_0' - m)$$

$$q = (1+A+am)^2 - 4am$$

$$p = P/K_0.$$

The 20°C isotherm reduced from the Hugoniot data of Christian (1957) assuming that the Grüneisen parameter, γ, is independent of temperature and volume is also shown. The curves marked with $+e$ and $-e$ indicate the estimated error limit for the experimental data of Hammond.

It is seen in Fig. 10 that the experimental data are consistent with the $P-V$ trajectory calculated from the Fritz–Thurston equation using the ultrasonic data of Barsch and Chang (1971) and with that reduced from the Hugoniot data. Since the experimental uncertainty in the static compression data of Hammond (1969) is relatively large, it is not possible to select one over the other on the basis of the static compression data. However, in view of the rather crude assumption, that γ is constant, used by Christian for the reduction of the Hugoniot data to the 20°C isotherm, and also of the precise nature of the ultrasonic measurements for K_0 and K'_0, the $P-V$ relation obtained using the Fritz–Thurston equation and the ultrasonic data of Barsch and Chang (1971) is considered to be preferable. The values for the Fritz–Thurston trajectory for CsI at 25°C are listed in Table 4.

3 Experimental results

3.1 METAL HALIDES

3.1.1. *Compression and phase transformations*

a. *Alkali halides*   Alkali halides have one of two crystal structures, the B1 (NaCl type) which is face centred cubic or the B2 (CsCl type) which is primitive cubic. In the B1 structure the cation is surrounded by six nearest neighbour anions while in the B2 structure the cation is surrounded by eight nearest neighbour anions. CsCl, CsBr, and CsI exist in the B2 phase at 1 bar and 25°C. All of the other alkali halides have the B1 structure at 445°C and 1 bar. Bridgman (1940, 1945) observed pressure-induced phase transitions in KCl, KBr, RbCl, RbBr, and RbI by dilatometry in a piston-cylinder apparatus. His measurements of pressure and volume change for each transition are given in Tables 5 and 6. The high-pressure phases for these salts were identified as B2 by Jacobs (1938), by Jamieson (1957), by Adams and Davis (1965), and by Weir and Piermarini (1964), and others.

Evidence for a phase transition in NaCl was found by Christian (1957), Alder (1963), and Kuzubov (1965) using shock methods. Fritz *et al.* (1971) also using shock loading, determined that the transformation takes place at 231 kbar and 1125 K. Bassett *et al.* (1968)

TABLE 5

Comparison between the experimental data obtained at room temperature by three separate investigations

	KCl		KBr		KI	
	$-\Delta V_t/V_{01}$ %	$-\Delta V_t$ cm³ mol⁻¹	$-\Delta V_t/V_{01}$ %	$-\Delta V_t$ cm³ mol⁻¹	$-\Delta V_t/V_{01}$ %	$-\Delta V_t$ cm³ mol⁻¹
Bridgman[a]	11.33 ± 0.33	4.25 ± 0.13	10.52 ± 0.30	4.51 ± 0.13	8.50 ± 0.25	4.52 ± 0.13
Weir and Piermarini[b]	20.2	7.60	18.9	8.18	15.3	8.14
Bassett et al. (1969)	11.0 ± 0.5	4.12 ± 0.18	10.6 ± 0.5	4.59 ± 0.25	8.4 ± 0.3	4.45 ± 0.16
Darnell and McCollum (1970)	11.0	4.12	9.6	4.15	8.3	4.41
Vaidya and Kennedy (1971)	11.22	4.21	10.27	4.44	8.50	4.51

[a] Bridgman (1940, 1945). The precision for Bridgman's data on $-\Delta V_t/V_{01}$ was estimated to be ± 3 per cent of a $-\Delta V_t/V_{01}$ value. This estimate is based on the precision of ± 1.5 per cent in a $-\Delta V/V_0$ for the compression data of Bridgman (1945); see his Table 1 on p. 3.

[b] Weir and Piermarini (1964); calculated from the X-ray diffraction data listed in their Table 1.

TABLE 6

The thermodynamic data for the B1–B2 phase transformations in eight alkali halides

	NaCl[a]	KCl	RbCl	CsCl	KBr	RbBr	KI	RbI
V_{01} (cm³ mol⁻¹)	26.997	37.47	43.09	50.32[c] (42.21)	43.21	48.78	53.12	59.54
Phase	B1	B1	B1	B1 (B2)	B1	B1	B1	B1
Transformation pressure (kbar)	300±10	19.7[b]	4.9[b]	0	18.1[b]	4.5[b]	17.8[b]	4.0[b]
Transformation temperature (K)	296	296	296	718	296	296	296	296
$-\Delta V_t$ at 296 K	1.1±0.1	4.25[b]	6.06[b]	8.85	4.51[b]	6.46[b]	4.52[b]	7.53[b]
$-\Delta V_t/V_{01}$ at 296 K	0.037	0.1133[b]	0.1406[b]	0.165[d]	0.1052[b]	0.1325[b]	0.0850[b]	0.1265[b]
dP/dT for the phase boundary (bar K⁻¹)	−83±10	−0.25[e] ±0.20	3.0[e] ±0.2	5.0[e] ±0.2	0.55[e] ±0.7	2.2[e] ±0.8	−1.9[e] ±0.6	1.94[e] ±0.40
$(\Delta S)_p$ (cal mol⁻¹ deg)	1.9 ±0.3	0.025 ±0.021	−0.44 ±0.04	−1.06 ±0.04	−0.06 ±0.07	−0.34 ±0.08	0.20 ±0.06	−0.35 ±0.07

[a] Bassett et al. (1968).

[b] Bridgman (1945).

[c] Calculated from the volume data at 728 K and 1 bar for an assumed volume thermal expansion coefficient of 1.5 × 10⁻⁴ K⁻¹.

[d] From the X-ray diffraction data of Swanson and Fuyat (1953) at 298 K and Wagner and Lippert (1936) at 728 K.

[e] Pistorius (1964, 1965a, 1965b).

observed a phase transformation in NaCl at 300 kbar and 25°C and identified the high-pressure phase as the B2 structure from X-ray diffraction photographs obtained with a diamond-anvil press. They calculated a volume change of 1.00 cm³ mol⁻¹ at the phase transformation. Earlier reports of a phase transformation in NaCl at pressures less than 50 kbar were apparently spurious as discussed by Johnson (1966) and by Bassett et al. (1968).

Bassett et al. (1969) measured the volume changes accompanying the phase transformations in KCl, KBr, and KI by X-ray diffraction. They undertook this investigation because of a disagreement between the values for the volume changes reported by Bridgman (1940, 1945) based upon dilatometric measurements and those of Weir and Piermarini (1964) based upon X-ray diffraction data obtained with a diamond press. By cycling the load on the sample, Bassett et al. (1969) were able to produce a fine mixture of the B1 and B2 phases. The volume change values which they obtained from this mixture agreed with Bridgman's values in Table 5. They concluded that the volume changes reported by Weir and Piermarini (1964) were too large because some of the B1 phase in their sample was at a pressure below the transition pressure while some of the B2 phase was at a pressure above the transition pressure, thereby exaggerating the apparent volume change. Weir and Piermarini's volume change measurements for the rubidium halides were also larger than those of Bridgman. Bassett et al. (1969) concluded that these were also subject to the same source of error and consequently favoured Bridgman's measurements for all of the potassium and rubidium halides. These are given along with other pertinent data in Table 6. Recent volume change measurements by Darnell and McCollum (1970) and Vaidya and Kennedy (1971) for the potassium halides are in good agreement with those of Bassett et al. (1969) as shown in Table 5.

Pistorius (1964, 1965a, 1965b) has measured the melting point as a function of pressure for several alkali halides. He was able to recognize a solid–solid phase transition from a change in the slope of the melting curve, thus establishing the pressure of a transition at an elevated temperature. Values of dP/dT for the B1–B2 transition in KCl, KBr, KI, RbCl, RbBr, and RbI based on Pistorius's measurements are given in Table 6. Bassett et al. (1968) obtained a value of -83 bar K⁻¹ for dP/dT of the B1–B2 boundary for NaCl on the basis of their own observation at 300 kbar and 298 K and of the transition value of 231 kbar at 1125 K observed by Fritz *et al.* (1971) in a single crystal of NaCl along the [111] direction under shock compression. The entropy changes associated with the B1–B2 transitions in the alkali halides were calculated from the volume data and the slopes of the boundaries using the Clapeyron equation, and are listed in Table 6.

A thermodynamic relationship between the entropy change and the volume change associated with a phase transformation at a constant temperature may be expressed by the equation derived by Bassett *et al.* (1968):

$$(\Delta S)_p \approx \gamma_{01} C v_1 (\Delta V_t / V_{01}) + (\Delta S)_v$$

where $(\Delta S)_p$ is the isobaric entropy change; $(\Delta S)_v$ is the isochoric entropy change or the change in the configurational entropies; γ_{01} is the Grüneisen parameter for the B1 phase at zero pressure; Cv_1 is the heat capacity for the B1 phase at zero pressure; ΔV_t is the volume change associated with the phase transformation at the transformation pressure; and V_{01} is the volume of the B1 phase at zero pressure. For these alkali halides the Debye temperature is nearly equal to or less than 25°C and they very closely obey the Dulong–Petit law. Therefore, they all have a heat capacity of approximately $6R$ (where R is the gas constant) at room temperature. The Grüneisen parameter is also nearly independent of composition and is approximately 1.6 ± 0.2 for these compounds. When these values are substituted into the equation, we obtain

$$(\Delta S)_p = (1.6 \pm 0.2)\ (12)\ (\Delta V_t / V_{01}) + (\Delta S)_v \text{ cal mol}^{-1} \text{ deg}^{-1}.$$

Since all of these alkali halides undergo the B1–B2 phase transformation, the change in the configurational entropy, $(\Delta S)_v$, may be assumed to be independent of the chemical composition and constant. In such a case the isobaric entropy change for the phase transformation is a linear function for the volume change associated with the phase transformation. A least-squares fit of the data to a linear equation (Fig. 11) yields

$$(\Delta S)_p = 19.0(\Delta V_t / V_{01}) + 2.3 \text{ (cal mol}^{-1} \text{ deg}^{-1}).$$

Thus the slope of the line is in good agreement with the predicted value of 19 ± 3 eu, and the entropy axis intercept is 2.3 eu. Therefore the difference between the configurational entropies between the B1 and B2 structures is 2.3 eu; the B2 structure has the greater entropy.

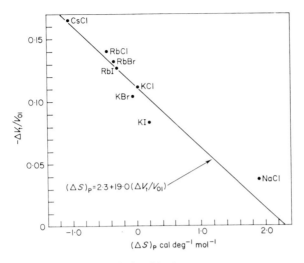

Fig. 11. The plot indicates the relationship between volume change and entropy change associated with the B1–B2 phase transformations in eight alkali halides. ΔV_t denotes the volume change associated with the phase transformation at transformation pressure and room temperature; V_{01}, the volume of the B1 phase at room temperature and zero pressure; and $(\Delta S)_p$, the entropy change asscoiated with the phase transformation at the transformation pressure and room temperature (Bassett et al., 1969).

b. *Silver halides*   It is of interest to compare the behaviour of silver halides with that of the alkali halides.

Polymorphism and compression of silver iodide has been the subject of a number of investigations. At room temperature and atomospheric pressure AgI consists of a mixture of wurtzite (B4) and sphalerite (B3) phases. Majumdar and Roy (1959) and Burley (1963, 1964) showed that the B4 phase is the thermodynamically stable one. At atmospheric pressure AgI transforms to a new phase having the B23 structure at 147°C and is stable up to 558°C, the melting temperature. This structure consists of a body-centred-cubic arrangement of anions with cations randomly distributed interstitially.

Bridgman (1915) studied the effect of pressure on volume of AgI up to 12 kbar and 200°C. He reported a volume discontinuity at 2.96 kbar and 30°C and proposed a triple point at 2.76 kbar and 99.4°C where phases I (B23), II (B3–B4), and III (B1) coexist. Bassett and Takahashi (1965) using visual observations in a diamond press equipped for heating the sample while under pressure placed this triple point at 3.1 ± 0.4 kbar and $98 \pm 2°C$. Jacobs (1938) identified the high-pressure phase as B1 by means of one of the first high-pressure X-ray diffraction cameras. Using a diamond anvil press, Van Valkenburg (1964) visually observed a strongly birefringent phase at about 3 kbar and 25°C separating the B3–B4 phases from the B1 phase. Bassett and Takahashi (1965) found that this phase terminates in a triple point at 65°C and 3.1 kbar. This birefringent phase has been the subject of X-ray diffraction studies by Davis and Adams (1964), Neuhaus and Hinze (1966), and Moore and Kasper (1968). From their diffraction data obtained with a diamond anvil press, Moore and Kasper (1968) deduced a tetragonal or pseudotetragonal unit cell containing two molecules and proposed an atomic arrangement resembling that of PbO.

Slykhouse and Drickamer (1958) using optical absorption and Riggleman and Drickamer (1963) using electrical resistance found a high-pressure phase of AgI stable above 97 kbar. Van Valkenburg (1964) found that this phase is birefringent. Bassett and Takahashi (1965) measured the pressure of the first appearance of this phase and found it to be 100 ± 5 kbar. They obtained an X-ray diffraction pattern of it but were unable to index the pattern. Schock and Jamieson (1969) obtained an X-ray diffraction pattern of this phase which was considerably different from the one obtained by Bassett and Takahashi (1965) but had five lines in common with theirs (Table 7). Schock and Jamieson point out that several of the strong lines reported by Bassett and Takahashi can be indexed as the B1 phase. Visual observations, however, indicate that the new phase is birefringent throughout. When Schock and Jamieson attempted to index their pattern they discovered that not only their lines, but nearly all of the lines of Bassett and Takahashi, including all of those which could be indexed as B1, fit a tetragonal indexing scheme. Schock and Jamieson conclude that the difference between the two patterns may be due to the different diffraction geometries employed by them and by Bassett and Takahashi. In their instrument the X-ray beam passes between tungsten carbide Bridgman anvils entering and leaving the sample through an anular gasket that is transparent to X-rays. In the diamond press used by Bassett and Takahashi the X-ray beam enters and leaves the sample through the diamond anvils. Thus, in the instrument used by Schock and Jamieson, the beam passes perpendicular to the axis of compression while in the diamond press the beam passes parallel to the axis of compression. If the Bridgman anvil geometry is causing preferential

TABLE 7

Proposed fit of X-ray data for the 100 kbar phase of AgI to a tetragonal cell with $a = 5.611$ Å and $c = 5.020$ Å (taken from Schock and Jamieson, 1969)

hkl	d_{calc} Å	$d_{obs}{}^a$ Å	$d_{obs}{}^b$ Å
200	2.805	2.827	2.879
022	2.510		
210	2.508	2.492	2.471
102	2.291	2.291 Ag(?)	2.337 Ag(?)
112	2.120	2.120	2.112
220	1.982	1.982	1.983
212	1.774⎤		
310	1.774 ⎬	1.764	
301	1.751⎦		
311	1.672⎤		1.668
003	1.669⎦		
302	1.500	1.503	
312	1.448		1.442
400	1.402	1.405	1.417
330	1.322	1.325	1.324
331	1.278⎤		1.293(?)
223	1.276⎦		
004	1.255⎤	1.256	
420	1.254⎦		
114	1.196⎤		
412	1.196 ⎬	1.184	1.179
332	1.170⎦		

[a] Schock and Jamieson (1969).
[b] Bassett and Takahashi (1965).

orientation of the samples in both cases, then this orientation would have a different effect in each instrument because of the difference in the direction of the X-ray beam. Schock and Jamieson point out that if 4 molecules are assigned to the tetragonal unit cell, the volume change at the transition would be 12.8 per cent, a value consistent with the 10 to 20 per cent predicted by Van Valkenburg (1964) from refractive index observations. Although the explanation put forth by Schock and Jamieson seems to be consistent with the observations, confirmation must await a study in which different geometries can be used on the same samples or a technique producing minimum orientation is used.

Bassett and Takahashi (1965) determined the compression of the B1 phase of AgI from 4 to 100 kbar. The zero-pressure volume (V_0) for the B1 phase cannot be determined directly because AgI reverts to other phases below 4 kbar. Therefore, the compression curve of the B1 phase (Fig. 12) was extrapolated to zero pressure, and a value of 34.09 $cm^3\ mol^{-1}$ was obtained. The P–V data for the B1 phase of AgI was fitted to the Birch–Murnaghan equation by least squares using this value for V_0 and assuming a value of 5.4 for K_0' in the Birch–Murnaghan

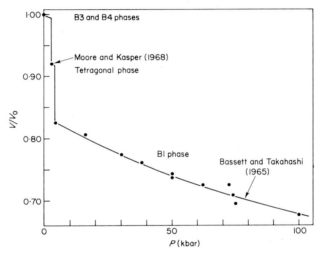

Fig. 12. Isothermal compression of AgI at 25°C calculated from lattice parameters determined by X-ray diffraction.

equation. This value for K_0' was assumed on the basis of the mean of the values for the alkali halides having the B1 structure. Fitting the P–V data to the Birch–Murnaghan equation yields a value of 260 ± 11 kbar for the zero-pressure bulk modulus (K_0).

Schock and Jamieson (1969) have shown that the B1 phases of AgCl and AgBr transform to the cinnabar (B9) structure (which is hexagonal) with pressure. Thus the B1 phases of silver halides, unlike the alkali halides, undergo pressure-induced phase transformations to phases which have lower symmetry than the B2 phase to which the alkali halides transform. This is probably due to the more covalent nature of the bonding in silver halides. The electronegativity difference in silver halides is much less than that in alkali halides. Their bonding is therefore more covalent and as a result the bonding is unequal in different directions.

3.2 METALS AND ALLOYS

Geophysical evidence suggests that the earth's core consists mainly of iron alloyed with nickel, silicon and other elements. Because of this current consensus regarding the earth's core, the effect of pressure on the volume and crystal structure of iron, iron-nickel and iron-silicon alloys was investigated. At high pressures, iron and its iron-rich alloys, which are in the bcc structure at low pressures, transform to the hcp structure. In order to extend our understanding of the behaviour of the hcp metals under pressure, the hcp phases of lead and rhenium were also investigated. The results of the studies are summarized below.

3.2.1 *Compression*

a. *Iron and its alloys* The experimental data for iron, Fe-5.15Ni, Fe-10.26Ni, Fe-7.2Si and Fe-25Si are presented in Figs 13, 14, and 15.

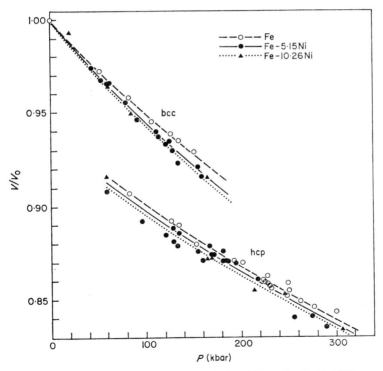

Fig. 13. Isothermal compression of Fe, Fe-5.15Ni, and Fe-10.26Ni at 25°C calculated from lattice parameters from X-ray diffraction (Takahashi *et al.*, 1968).

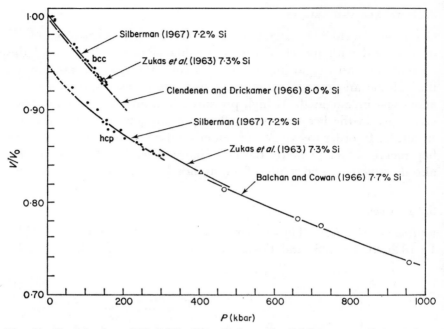

Fig. 14. Compression of Fe-7.2Si. The circles and solid lines are calculated from lattice parameters determined by X-ray diffraction at 25°C (Silberman, 1967) These measurements are compared with the shock Hugoniots of Zukas *et al.* (1963) and Balchan and Cowan (1966).

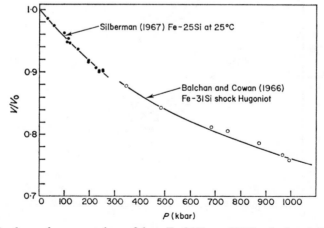

Fig. 15. Isothermal compression of bcc Fe-25Si at 25°C calculated from lattice parameters from X-ray diffraction (Silberman, 1967) with the shock Hugoniot for Fe-31Si (Balchan and Cowan, 1966).

The data are also compared with the shock Hugoniot data of McQueen and Marsh (1960, 1966) for iron and iron-nickel alloys, and of Balchan and Cowan (1966) and Zukas et al. (1963) for iron-silicon alloys. The bulk modulus at zero pressure (K_0) and its pressure derivative (K_0'), $(\partial K_0/\partial P)_T$, were calculated by means of the least-squares fit of the P–V data to the Birch–Murnaghan equation:

$$P = 1.5 K_0 [V/V_0)^{-7/3} - (V/V_0)^{-5/3}] \{ \tfrac{3}{4}(4 - K_0')[(V/V_0)^{-2/3} - 1] \},$$

where V and V_0 are respectively the molar volumes at pressures P and zero. Since iron and its alloys exist in the bcc structure at 1 bar, the experimental values for V_0 are available. In those cases, the values for K_0 and K_0' can be obtained with a reasonable precision by the least-squares method. However, in the cases of the high-pressure hcp phases, the volume at zero pressure cannot be determined experimentally since the high-pressure phase reverts to the low-pressure bcc phase upon reducing the pressure. Therefore, for the high-pressure phases, the K_0' value was assumed to be 4.0, and the values of K_0 and V_0 were obtained as the least-squares solutions. The values for V_0, K_0, β_0 and K_0' thus calculated are listed in Table 8.

Takahashi and Liu (1970) have shown that the zero-pressure compressibility (β_0) of binary solid solutions is proportional to the mole fraction of a component. In Fig. 16, the compressibilities for the bcc phases of iron, iron-nickel and iron-silicon alloys are plotted against the mole fraction of the alloying elements. It appears that the compressibilities of those alloys are a linear function of the mole fraction of nickel and silicon, and that the addition of nickel causes an increase of compressibility whereas silicon causes a decrease of compressibility of the alloys. Although the effect of the nickel on the compressibility is opposite to that of silicon, the magnitude of the effect appears to be similar.

b. *Lead*   The compression data for lead were obtained by Mao et al. (in preparation) using pure iron, NaCl and CsI as pressure standards, and the results are shown in Fig. 17. The data are also compared with the shock compression Hugoniot data of McQueen and Marsh (1960) and Alt'schuler et al. (1960) and with the room-temperature data of Bridgman (1945). The phase transformation reported by Balchan and Drickamer (1961) was identified by Takahashi et al. (1969) as a fcc to hcp transformation by means of the X-ray diffraction method. The pressure of transformation was also determined to be 130 ± 10 kbar

TABLE 8

The values of V_0, K_0, β_0 and $(\partial K_0/\partial P)_T$ calculated from the experimental data for various iron alloys, lead and rhenium

	V_0 $\text{cm}^3\,\text{mol}^{-1}$	K_0 Mbar	β_0 Mbar^{-1}	K'_0	References
The bcc phase					
Fe	7.093	1.639	0.610	5.94	Rotter and Smith (1966)
Fe-5.2Ni	7.110	1.62 ± 0.05	0.62 ± 0.02	5.5 ∓ 0.8	Takahashi et al. (1968)
Fe-10.3Ni	7.117	1.56 ± 0.07	0.64 ± 0.03	4.2 ∓ 0.8	Takahashi et al. (1968)
Fe-7.2Si	7.058	1.53 ± 0.07	0.65 ± 0.03	5.7 ∓ 0.8	Takahashi et al. (1968)
Fe-25Si	6.797	1.75 ± 0.08	0.57 ± 0.03	4.3 ∓ 1.0	Silberman (1967)
		2.14 ± 0.09	0.47 ± 0.02	3.5 ∓ 0.8	Silberman (1967)
The fcc phase					
Pb	18.269	0.437 ± 0.013	2.30 ± 0.06	4.4 ∓ 0.4	Mao et al. (in preparation)
The hcp phase					
Fe	6.68 ∓ 0.02	2.08 ± 0.10	0.49 ± 0.03	4.0[a]	Takahashi et al. (1968)
Fe-5.2Ni	6.66 ∓ 0.02	2.12 ± 0.15	0.48 ± 0.03	4.0[a]	Takahashi et al. (1968)
Fe-10.3Ni	6.67 ∓ 0.02	2.15 ± 0.25	0.47 ± 0.03	4.0[a]	Takahashi et al. (1968)
Fe-7.2Si	6.72 ∓ 0.03	1.88 ± 0.14	0.53 ± 0.03	4.0[a]	Silberman (1967)
Re	8.86	3.50 ± 0.15	0.29 ± 0.02	4.3 ∓ 1.0	Liu et al. (1970)

The heading of each column is defined as follows:

$$V_0 = V|_{p=0}; \quad K_0 = -V(\partial P/\partial V)_T|_{p=0};$$
$$\beta_0 = 1/K_0;$$
$$K'_0 = (\partial K/\partial P)_T|_{p=0}.$$

[a] The value for K'_0 was fixed at 4.0, and the values for V_0 and K_0 were determined by the least-squares fit of the data to the Birch–Murnaghan equation.

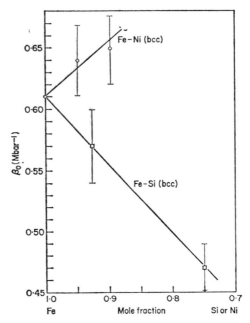

Fig. 16. The composition dependence of the compressibility of the bcc phases in the iron-nickel and iron-silicon alloys.

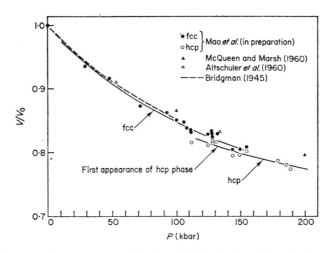

Fig. 17. Isothermal compression of lead at 25°C calculated from lattice parameters from X-ray diffraction compared with measurements by other techniques.

at room temperature (Takahashi *et al.*, 1969). The least-squares fit of the room-temperature P–V data to the Birch–Murnaghan equation yields the K_0 value of 437 ± 13 kbar, and the K_0' value of 4.4 ∓ 0.4 kbar. On the other hand, using an ultrasonic pulse-echo method, Miller and Schuele (1969) determined the isothermal bulk modulus (K_0) and its pressure derivative (K_0') at 293 K to be 419 ± 2 and 5.7 kbar respectively. Thus the K_0 values of Miller and Schuele are approximately 4 per cent smaller than the K_0 value of Mao *et al.* However, if the upper limit of the value of Miller and Schuele is compared with the lower limit of the value of Mao *et al.*, the difference is only 0.7 per cent.

Since the sonic determination of K_0 is more reliable than the calculation from the P–V relation, the K_0 value of Miller and Schuele was accepted as the correct value, and the K_0' value was computed from the P–V data by the least-squares method. For the value of $K_0 = 419$ kbar, a K_0' value of 4.8 was obtained. This value of K_0' is about 20 per cent smaller than the K_0' value of 5.7 obtained by the sonic means. However, the K_0' value obtained by fitting the P–V data represents an average over the pressure range in which the data were obtained. As pointed out by Anderson (1967), the value of K' decreases with pressure and, therefore, the value of K_0' obtained by fitting the P–V data should be smaller than the K_0' value obtained by the sonic method which yields the K' value at zero pressure rather than an average over a pressure

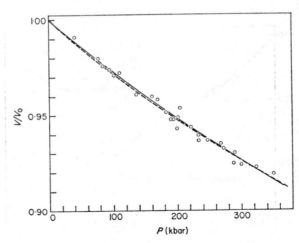

Fig. 18. Isothermal compression of rhenium at 25°C calculated from lattice parameters determined by X-ray diffraction (Liu *et al.*, 1970). The two curves indicate respectively the results of different fitting procedures of the P–V data.

range. Therefore, the difference between the values of the pressure derivative of bulk modulus is partially due to the ways in which these values were obtained.

c. *Rhenium*   Rhenium exists in the hcp structure up to 350 kbar, the maximum pressure achieved in the investigation. The observed *P–V* relation obtained by X-ray diffraction is shown in Fig. 18, and the values of $K_0 = 3.50 \pm 0.15$ and $K'_0 = 4.3 \mp 1.0$ Mbar were obtained by means of the least-squares fit of the experimental data to the Birch–Murnaghan equation (Liu *et al.*, 1970). The value of K_0 thus determined is in good agreement with the sonic value of 3.587 Mbar determined by Shepard and Smith (1965), and also in agreement with the value of 3.69 ± 0.1 Mbar obtained by Bridgman (1955) by his dilatometric method.

3.2.2   *Phase transformations*

a. *Iron and its nickel and silicon alloys*   Evidence for a phase transformation in iron at 130 kbar was first recognized by Bancroft *et al.* (1956). The structure of the high-pressure phase was identified as hexagonal close-packed by Takahashi and Bassett (1964) and Clendenen and Drickamer (1964). The lattice parameters of both the low-pressure phase (bcc) and the high-pressure (hcp) have been measured as a function of pressure by Mao *et al.* (1967), who reported a volume change at the transition of -0.34 ± 0.01 cm³ mol⁻¹.

In the pressure–temperature phase diagram of iron (Fig. 19), the boundaries separating the bcc, hcp and fcc phases meet at a triple point at approximately 500°C and 110 kbar (Bundy, 1965). Mao *et al.* (1967) were able to show that their value for the bcc–hcp volume change is consistent with the Clapeyron equation,

$$\frac{\mathrm{d}T}{\mathrm{d}P} = \frac{\Delta V}{\Delta S},$$

the thermodynamic requirement of a triple point,

$$\Sigma \Delta V = 0, \qquad \Sigma \Delta S = 0$$

and previously published values for volume changes at the phase boundaries. These values are summarized in Table 9.

Takahashi *et al.* (1968) reported that the addition of 10 atomic per cent of nickel results in an increase of the volume change associated

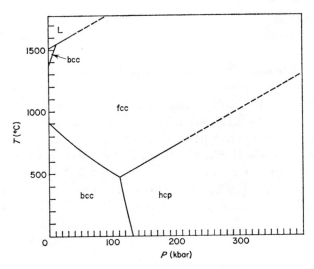

Fig. 19. The phase diagram of iron.

with the bcc–hcp transition from -0.34 ± 0.02 cm³ mol⁻¹ to $-0.36 \pm$ 0.02 cm³ mol⁻¹ (Table 10). Silberman (1967) investigated the effect of silicon on the phase transformation in iron by X-ray diffraction. He found that the presence of 7.2 atomic per cent of silicon results in a decrease in the volume change associated with the bcc–hcp transformation from -0.34 ± 0.02 to -0.31 ± 0.02 cm³ mol⁻¹ (Table 10).

This decrease in the volume change associated with the phase transition seems to be due entirely to the effect of silicon on the bcc phase since it has little effect on the hcp phase (Fig. 20). Farquhar *et al.* (1945) have shown that the unit cell dimension of the bcc phase of

TABLE 9

The volume changes at the phase boundaries around the triple point in the iron phase diagram

Phase boundary	ΔV cm³ mol⁻¹	References
bcc–hcp	−0.34	Mao *et al.* (1967)
hcp–fcc	+0.13	Blackburn *et al.* (1965)
fcc–bcc	+0.21	Blackburn *et al.* (1965)
		Kaufman *et al.* (1963)

TABLE 10

Volume change associated with phase transformation in metals

Sample	Volume change at transition cm³ mol⁻¹	References
Fe	-0.34 ± 0.02	Takahashi *et al.* (1968)
Fe-5Ni	-0.36 ± 0.02	Takahashi *et al.* (1968)
Fe-10Ni	-0.36 ± 0.02	Takahashi *et al.* (1968)
Fe-7Si	-0.31 ± 0.02	Silberman (1967)
Pb	-0.18 ± 0.06	Takahashi *et al.* (1969)

iron-silicon alloys at atmospheric pressure and room temperature decreases with increasing silicon content. At 10 per cent silicon, initial ordering of the silicon atoms takes place to produce a super-lattice. At 25 per cent, the ordering is complete and the resulting phase is an intermetallic compound (Fe_3Si) having the BiF_3 structure shown in Fig. 21. In this structure the silicon atom is surrounded by eight iron atoms. The sp^3 nature of the silicon favours this position since alternate iron atoms of the nearest neighbour atoms are at apices of a tetrahedron. The greater covalent nature of the sp^3 electron orbitals in silicon even at concentrations below 10 per cent is probably responsible for the

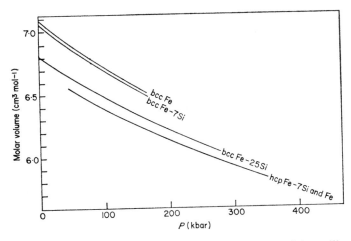

Fig. 20. Collected compression curves for the phases of iron and iron-silicon alloys.

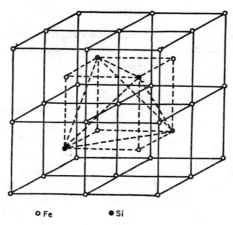

o Fe   • Si

Fig. 21. The structure of Fe$_3$Si.

decrease of the unit cell dimension with increase of silicon content in the bcc phase of iron-silicon alloys. Silberman's data for the compression of the Fe-7.2Si alloy show a decrease of approximately 0.45 per cent in the molar volume of the bcc phase over the whole pressure range over which the bcc phase is stable (Fig. 20). His measurements of the hcp phase, however, reveal no decrease in the molar volume of this phase at any pressure between 60 and 238 kbar due to the presence of 7.2 per cent silicon. These observations are consistent with the sp^3 configuration of the silicon atoms since in the hcp structure each atom is surrounded by 12 nearest neighbours. The tetrahedral geometry of the sp^3 configuration does not conform to the 12-fold site and should, therefore, have little of the influence on interatomic distances that it had in the 8-fold position of the bcc structure. Thus the substitution of silicon for iron in the hexagonal close-packed structure should have a negligible effect on the molar volume of that phase.

This is further substantiated by the shock-wave measurements of Zukas *et al.* (1963) on seven iron-silicon alloys ranging up to 13 per cent silicon. These measurements show that the silicon content has essentially no effect on the molar volume of the high-pressure (presumably hcp) phase.

As discussed in an earlier section and illustrated in Fig. 14, the compression measurements on the Fe-7.2Si alloy made by Silberman (1967) are in good agreement with the high-pressure X-ray diffraction measurements made by Clendenen and Drickamer (1966) on the bcc phase of an Fe-8.0Si alloy, and with the shock Hugoniot data of Zukas *et al.* (1963) and Balchan and Cowan (1966) on alloys between 7 and 8 atomic per cent silicon. Thus the features of the phase transformation discussed here are in agreement with the results of other investigations on alloys of similar composition.

Silberman (1967) found that the Fe-25Si alloy does not undergo a pressure-induced phase transformation at 25°C up to 238 kbar, the maximum pressure employed in his study of this alloy. As discussed in an earlier section, Silberman's static compression measurements when extrapolated to higher pressures are in agreement with the shock compression measurements of Balchan and Cowan (1966) on a sample of Fe-31Si (Fig. 15). It would appear, therefore, that the Fe-25Si alloy does not undergo a phase transformation up to a pressure of 1000 kbar, the maximum to which Balchan and Cowan subjected their specimen.

Silberman's data also show that the presence of the 25 atomic per cent silicon results in a 4.1 per cent decrease in molar volume at atmospheric pressure and a 3.4 per cent decrease in molar volume at 150 kbar (Fig. 20). In the absence of the hcp phase of the Fe-25Si alloy we can only assume that its molar volume would be similar to that of pure iron. This assumption seems justified in view of the superposition of the compression curves for the hcp phase of pure iron and for Silberman's 7.2 per cent silicon alloy of iron (Fig. 20) and the similar molar volume data of Zukas *et al.* (1963) for the high-pressure phase of seven alloys (up to 13 per cent silicon) described above. The volume change that would accompany a bcc–hcp transformation in Fe-25Si, if it were to take place, would be 0.10 cm^3 mol^{-1} at 150 kbar or less than a third of the volume change in pure iron.

Evidence for the phase transformation pressure for iron and each of its alloys comes from the X-ray diffraction studies as well as shock measurements. In Table 11 information pertaining to the pressure of transformation from three different sources is given: (1) the pressure at which the high-pressure phase first appears as load is being increased; (2) the pressure at which the high-pressure phase disappears as load is being decreased; (3) the pressure of first appearance of the high-pressure phase during shock loading. In our X-ray diffraction studies, the determination of transformation pressure was not a major objective. In some cases the spacing of the data points in the vicinity of the transformation is sufficient only to bracket the pressure at which the onset or reversal

TABLE 11

The observed phase transformation pressure under different experimental conditions

	Low-pressure phase	High-pressure phase	$P_{onset}{}^a$ kbar	$P_{rev}{}^b$ kbar	$P_{shock}{}^c$ kbar	References
Fe	bcc	hcp	$107 < P < 127$	$P \approx 83$	130	Takahashi et al. (1968) Bancroft et al. (1956)
Fe-5.2Ni	bcc	hcp	$126 < P < 129$	$P \approx 60$	122	Takahashi et al. (1968) Fowler et al. (1961)
Fe-10.3Ni	bcc	hcp	$85 < P < 166$	$0 < P < 60$	118	Takahashi et al. (1968) Fowler et al. (1961)
Fe-7.2Si	bcc	hcp	$149 < P < 159$	$0 < P < 65$	150	Silberman (1967) Zukas et al. (1963)
Pb	fcc	hcp	$128 < P < 131$	$0 < P < 100$	—	Mao et al. (1969b)

[a] Pressure at which the low-pressure phase starts to transform to the high-pressure phase when load is increased.
[b] Pressure at which the high-pressure phase disappears when the load is reduced.
[c] Pressure at which the phase transformation starts under shock loading.

of the transformation took place. The X-ray data by themselves are not adequate to show the effect of nickel concentration on the transformation pressure. The agreement between the pressures bracketed by the X-ray diffraction measurements and those based upon the shock data is sufficiently good that there is no reason to question the decrease of transformation pressure with increasing nickel content reported by Fowler *et al.* (1961) based on the shock data. Both the X-ray diffraction and the shock measurements show a substantial increase in transformation pressure with increasing silicon concentration.

The pressure at which the last of the high-pressure phase is observed when the load is released is of interest as an indicator of the order of magnitude of the hysteresis involved. Although the thermodynamic equilibrium transformation pressure cannot be determined by the present X-ray method, it most likely lies between the first appearance with increasing pressure and the last appearance with decreasing pressure.

Some interesting studies of the pressure of the phase transformation in iron have been made by comparing iron and lead in the same pressure device (Mao *et al.*, 1969b; Vereshchagin *et al.*, 1969). Discussion of these studies is deferred to section 4.

b. *Lead* Balchan and Drickamer (1961) reported a 23 per cent increase in the electrical resistance of lead at 161 kbar and 25°C. Klement (1963) predicted that the structure of the high-pressure phase is hcp from his studies of a similar transition in bismuth-lead alloys. Takahashi *et al.* (1969) obtained X-ray diffraction photographs of the high-pressure phase of lead from which they were able to identify the phase as the hexagonal-close-packed structure. They reported a volume change for the fcc–hcp transformation of -0.18 ± 0.06 cm^3 mol^{-1}. They observed the first appearance of the most intense lines of the high-pressure phase in the diffraction pattern made with the diamond anvil press at a pressure between 128 and 131 kbar (Table 11). In section 4 of this review, this pressure of the phase transformation in lead is compared with determinations made in other types of apparatus.

3.2.3 *Pressure dependence of axial ratio in metals*

a. *Iron and its nickel and silicon alloys* The effect of pressure on the c/a ratio has been determined for hexagonal-close-packed phases of Fe-5Ni, Fe-10Ni, Fe-7Si, Pb and Re, and the effect of temperature on the c/a

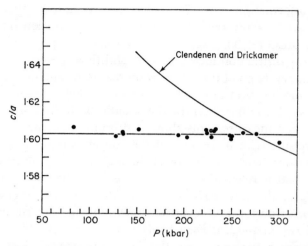

Fig. 22. The axial ratio (c/a) of hcp iron as a function of pressure at 25°C determined by X-ray diffraction (Takahashi *et al.*, 1968) compared with the determination by Clendenen and Drickamer (1964).

ratio of Re has been determined. The results of these investigations are shown in Figs 22–27. For all of these substances the c/a ratio is virtually independent of pressure. Even though the c/a ratio for Re is independent of pressure, it shows a dependence on temperature.

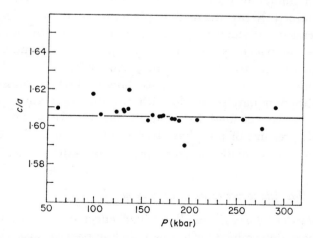

Fig. 23. The axial ratio (c/a) in the hcp phase of Fe-5.15Ni as a function of pressure at 25°C determined by X-ray diffraction (Takahashi *et al.*, 1968).

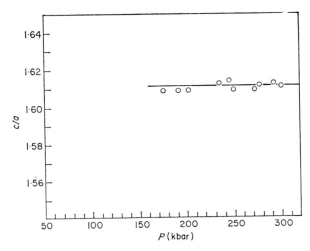

Fig. 24. The axial ratio (c/a) in the hcp phase of Fe-10.26Ni as a function of pressure at 25°C determined by X-ray diffraction (Takahashi *et al.*, 1968).

Mao *et al.* (1967) determined the c/a ratio for the hcp phase of pure iron to be 1.603 ± 0.001 over the pressure range from 80 to 310 kbar. This is in contrast with the findings of Clendenen and Drickamer (1964) who reported that the c/a ratio in hcp iron decreases from 1.645 to 1.580 with increasing pressure from 150 to 400 kbar. Mao *et al.* (1967)

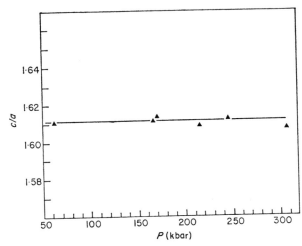

Fig. 25. The axial ratio (c/a) in the hcp phase of Fe-7.2Si as a function of pressure at 25°C determined by X-ray diffraction (Silberman, 1967).

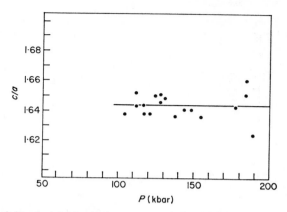

Fig. 26. The axial ratio (c/a) in the hcp phase of lead as a function of pressure at 25°C determined by X-ray diffraction (Mao *et al.*, in preparation).

were unable to explain the discrepancy between their observations and those of Clendenen and Drickamer (1964). Subsequently, however, Drickamer (personal communication, 1966) has reported that a new interpretation of their 1964 data yields a c/a ratio for iron that is nearly independent of pressure: 1.61 at 200 kbar and 1.595–1.60 at 400 kbar.

Takahashi *et al.* (1968) report an axial ratio of $c/a = 1.606 \pm 0.001$ for Fe-5Ni over the pressure range from 60 to 290 kbar. For the Fe-10Ni alloy they report $c/a = 1.611 \pm 0.001$ over the pressure range from 61 to 308 kbar. Thus there is a significant relationship between axial ratio

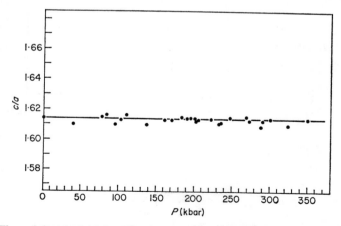

Fig. 27. The axial ratio (c/a) in rhenium, which has the hcp structure, as a function of pressure at 25°C (Liu *et al.*, 1970).

and nickel content in the iron-nickel alloys which Takahashi *et al.* (1968) express by $c/a = 1.603 + 0.08$ (Ni atomic per cent).

Lattice parameter determinations on the hcp phase of the Fe-7Si alloy by Silberman (1967) show that the c/a ratio is 1.611 ± 0.001 over the pressure range from 176 to 305 kbar. At pressures smaller than 176 kbar the (110) reflection of the bcc phase interferes with the (002) reflection of the hcp phase. Therefore, the axial ratio measurements made in the pressure range where the bcc phase coexists with the hcp phase were discarded.

b. *Lead*   Takahashi *et al.* (1969) identified the high-pressure phase of lead as hexagonal close-packed and reported an axial ratio of $c/a = 1.650 \pm 0.001$. This value is constant over the pressure range from 100 to 190 kbar as shown in Fig. 26 (Mao *et al.*, 1969b).

c. *Rhenium*   Rhenium has been investigated by Liu *et al.* (1970). From their measurement of lattice parameters as a function of pressure and temperature, they determined that the axial ratio has a value of 1.613 ± 0.002 and is constant up to 350 kbar. The axial ratio, however, varies with temperature. The coefficient of thermal expansion along the c-axis in Re is $(4.96 \pm 0.07) \times 10^{-6}°C^{-1}$ and along the a-axis it is $(7.03 \pm 0.10) \times 10^{-6}°C^{-1}$. Thus, the c-axis thermal expansion is 0.7 times as great as the a-axis thermal expansion. The difference between the effect of temperature and of pressure on the axial ratio can be explained in terms of lattice vibrations. The thermal expansivity is a manifestation of the anharmonicity in lattice vibrations, whereas the compressibility is primarily a manifestation of harmonic properties in solids. Hence, the most probable explanation for the temperature dependence of the c/a ratio in Re is that it depends mainly on a difference of anharmonicity in different directions.

d. *Comparison with other metals*   McWhan (1965) investigated the effect of pressure on the lattice parameters of Cd and Zn which have hexagonal close-packed structures. He reported that the linear compressibility of the c-axis in Cd is four times as great as the a-axis compressibility. In Zn the c-axis compressibility is six times as great as the a-axis compressibility. On the other hand, the thermal expansivity in Cd is nearly three times as much along the c-axis as along the a-axis, and in Zn it is more than four times as much along the c-axis as along the a-axis. Thus, the behaviour of these two metals seems to be quite different from the metals which we have studied. The c/a ratio of Cd is 1.886 and for Zn is 1.856. The metals which we have studied, on the other hand, have

c/a ratios ranging from 1.603 to 1.650, which are much closer to the ideal axial ratio of 1.633 for the hexagonal close-packing of hard spheres. Thus, it appears to be that in metals with axial ratios substantially different from the ideal ratio, pressure tends to make interatomic distances more nearly equal, thus changing the axial ratio towards the ideal ratio.

3.3 OXIDES AND SILICATES

3.3.1 *Compression*

Since the earth's mantle (33 to 2900 km deep) is presumably composed of oxides of such elements as silicon, magnesium, iron and aluminium, it is of geophysical interest to understand the effect of pressure on the lattice parameters and volumes of these compounds. Various ferromagnesian oxides and silicates have been examined at room temperature up to the maximum pressure of 300 kbar, which is equivalent to the pressure prevailing at a depth of about 1000 km.

The crystalline species studied thus far by our group are listed in Table 12. Among these fifteen samples, four of them (wüstite, haematite, ilmenite and magnetite) can be formed under low-pressure conditions, whereas eleven of them can be formed only under high-pressure conditions. The experimental P–V data are shown in Figs 28–38. Since the compression data for MgO has already been discussed in conjunction with the NaCl pressure scale (section 2.2.1), it is omitted here.

a. *Bulk moduli* The bulk modulus of each of these samples was computed by least-squares fit to the Birch–Murnaghan equation (see section 3.2.1). and the values are listed in Table 12. For all the samples, the value of K_0 was obtained for the value of K_0' set equal to 4.0. Since the K_0' values for the SiO_2 polymorphs, α-quartz and stishovite, have a value of 6.4 for the former (Soga, 1968a) and 7 for the latter (Ahrens *et al.*, 1970), the values of K_0 for coesite and stishovite were also calculated with K_0' set at 6.0 and 7.0. For the garnet group minerals, the K_0' value of 5.45 determined by Soga (1967) for an almanditic garnet was used throughout. When the precision of the data is good enough for calculating both K_0 and K_0', these two parameters were obtained as the least-squares solutions. The K_0' values thus obtained can be identified in Table 12 by the $\mp$ signs attached to them.

In order to test the reliability of the bulk modulus values thus

TABLE 12

Crystallographic and elastic properties for various oxides and silicates

Crystalline species	Crystallographic system	Unit cell dimension[a] Å	Molar volume[a] cm³ mol⁻¹	Density[a] g cm⁻³	Mean atomic weight g per atom	K_0 Mbar	K_0' $(\partial K/\partial P)_t$	References
Wustite $Fe_{0.924}O$	cubic (B1)	4.289 ± 0.002	12.04	5.745	34.61	1.42 ± 0.10	4.0	Mao et al. (1969c)
Haematite Fe_2O_3	hexagonal	$a = 5.038 \pm 0.002$ $c = 13.757 \pm 0.005$	30.35	5.262	31.94	2.28 ± 0.15 2.22 ± 0.15	4.0 4.53^b	Bassett et al. (in preparation)
Ilmenite $(Mg, Fe)TiO_3$	hexagonal	$a = 5.075 \pm 0.002$ $c = 13.972 \pm 0.005$	37.28	4.465	27.93	1.77 ± 0.07 1.77 ± 0.13	4.0 4.0 ∓ 1.3^c	Liu et al. (in preparation)
Magnetite Fe_3O_4	cubic (spinel)	8.395	44.53	5.200	33.08	1.83 ± 0.06	4.0	Mao (1967)
$\gamma\text{-}Fe_2SiO_4$	cubic (spinel)	8.234 ± 0.003	42.03	4.849	29.11	2.12 ± 0.10	4.0	Mao et al. (1969c)
$\gamma\text{-}(Fe_{0.9}Mg_{0.1})_2SiO_4$	cubic (spinel)	8.223 ± 0.003	41.88	4.718	28.21	1.96 ± 0.10	4.0	Mao et al. (1969c)
$\gamma\text{-}(Fe_{0.8}Mg_{0.2})_2SiO_4$	cubic (spinel)	8.204 ± 0.003	41.58	4.599	27.31	2.08 ± 0.10	4.0	Mao et al. (1969c)
$\gamma\text{-}Co_2SiO_4$	cubic (spinel)	8.139 ± 0.003	40.58	5.173	29.99	2.10 ± 0.06 2.19 ± 0.12	3.1 ∓ 0.6^a 4.0	Liu et al. (in preparation)

TABLE 12 (*Cont.*)

Crystalline species	Crystallographic system	Unit cell dimension[a] Å	Molar volume[a] cm³ mol⁻¹	Density[a] g cm⁻³	Mean atomic weight g per atom	K_0 Mbar	K_0' $(\partial K/\partial P)_t$	References
γ-Ni$_2$SiO$_4$	cubic (spinel)	8.043 ± 0.003	39.17	5.348	29.93	2.14 ± 0.07	4.0	Mao *et al.* (1970)
Coesite SiO$_2$	monoclinic	$a = 7.143 \pm 0.007$ $c = 12.36 \pm 0.01$ $\gamma = 120°$	20.56	2.925	20.03	1.26 ± 0.20 1.19 ± 0.02	4.0 6.0	Bassett and Barnett (1970)
Stishovite SiO$_2$	tetragonal	$a = 4.180 \pm 0.002$ $c = 2.666 \pm 0.001$	14.03	4.285	20.03	3.47 ± 0.08 3.35 ± 0.13	4.0 5.7 ± 1.7[c]	Liu *et al.* (in preparation)
	tetragonal	$a = 4.178 \pm 0.001$ $c = 2.663 \pm 0.001$	14.00	4.291	20.03	2.96 ± 0.25 2.85 ± 0.25	4.0 7.0	Bassett and Barnett (1970)
Pyrope Mg$_3$Al$_2$Si$_3$O$_{12}$	cubic (garnet)	11.454 ± 0.003	113.12	3.564	20.16	2.05 ± 0.07 1.90 ± 0.06	4.0 5.45	Takahashi and Liu (1970)
Pyrope-almandite (Mg$_{0.6}$Fe$_{0.3}$)$_3$Al$_2$Si$_3$O$_{12}$	cubic (garnet)	11.521 ± 0.003	115.12	3.81	21.93	1.93 ± 0.06 1.77 ± 0.06	4.0 5.45[c]	Takahashi and Liu (1970)
Almandite-pyrope (Mg$_{0.2}$Fe$_{0.8}$)$_3$Al$_2$Si$_3$O$_{12}$	cubic (garnet)	11.523 ± 0.003	115.17	4.160	23.79	1.86 ± 0.05 1.73 ± 0.06	4.0 5.45[c]	Takahashi and Liu (1970)
Almandite Fe$_3$Al$_2$Si$_3$O$_{12}$	cubic (garnet)	11.532 ± 0.003	115.45	4.312	24.89	1.80 ± 0.05 1.68 ± 0.05	4.0 5.45[c]	Takahashi and Liu (1970)

[a] Quantities determined at 1 bar and 23±2°C.

[b] $(\partial K/\partial P)_t$ based on sonic velocity measurements (Liebermann and Schreiber, 1968).

[c] The values for the zero-pressure bulk modulus and its pressure derivative were calculated by the least squares fit to the two-parameter Birch–Murnaghan equation.

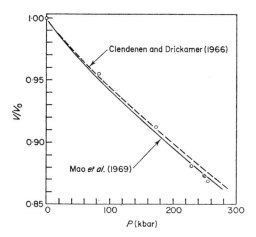

Fig. 28. Isothermal compression of Wüstite ($Fe_{0.924}O$) at 25°C calculated from lattice parameters determined by X-ray diffraction (Mao *et al.*, 1969c).

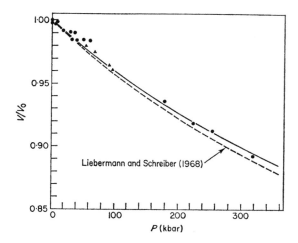

Fig. 29. Isothermal compression of haematite (Fe_2O_3) at 25°C calculated from lattice parameters determined by X-ray diffraction (Bassett *et al.*, in preparation) compared with the Murnaghan extrapolation of sonic velocity data of Liebermann and Schreiber (1968). The triangles represent data points obtained with a tetrahedral press and the circles those obtained with a diamond press.

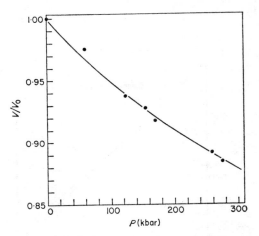

Fig. 30. Isothermal compression of magnesian ilmenite at 25°C calculated from lattice parameters determined by X-ray diffraction (Liu *et al.*, in preparation).

obtained, a portion of the same sample of the garnet, $Pyrope_{22}Almandite_{78}$, used for the single-crystal sonic determination by Soga (1967), was studied by the X-ray diffraction method using the diamond anvil press (Takahashi and Liu, 1970). The adiabatic bulk modulus obtained by Soga (1967) is 1.770 Mbar at 25°C and is in agreement with the result of Takahashi and Liu (1970), 1.74 ± 0.06 Mbar, which was

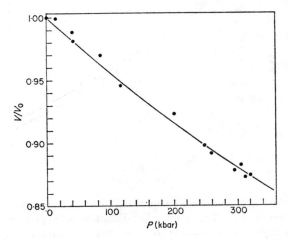

Fig. 31. Isothermal compression of magnetite (Fe_3O_4) at 25°C calculated from lattice parameters determined by X-ray diffraction (Mao, 1967).

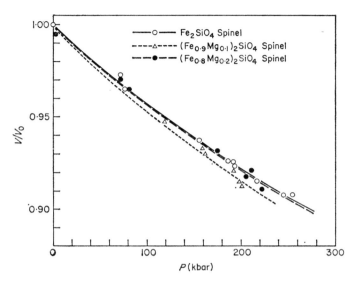

Fig. 32. Isothermal compression of three spinel phases of $(Mg, Fe)_2SiO_4$ at 25°C calculated from lattice parameters determined by X-ray diffraction (Mao *et al.*, 1969c).

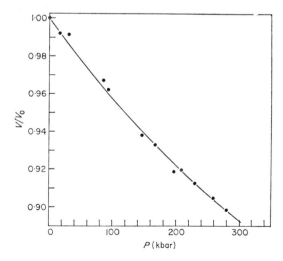

Fig. 33. Isothermal compression of the spinel phase of Co_2SiO_4 at 25°C calculated from lattice parameters determined by X-ray diffraction (Liu *et al.*, in preparation).

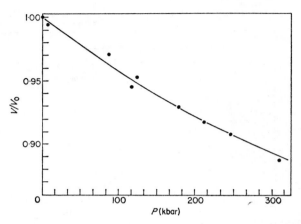

Fig. 34. Isothermal compression of the spinel phase of Ni_2SiO_4 at 25°C calculated from lattice parameters determined by X-ray diffraction (Mao *et al.*, 1970).

converted to the adiabatic value using the values of 1.5 for the Grüneisen parameter and 1.57×10^{-5} K^{-1} for the volume thermal expansion.

Both the tetrahedral press (Barnett and Hall, 1964) and the diamond press (Bassett *et al.*, 1967) have been used to obtain compression data for haematite (Fig. 29) and stishovite (Fig. 36). In the case of haematite, both sets of data are in good agreement and have been combined to determine the bulk modulus. In the case of stishovite, separate bulk moduli have been calculated from the studies of Bassett and Barnett

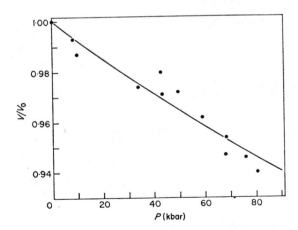

Fig. 35. Isothermal compression of coesite at 25°C calculated from lattice parameters determined by X-ray diffraction using a tetrahedral press (Bassett and Barnett, 1970).

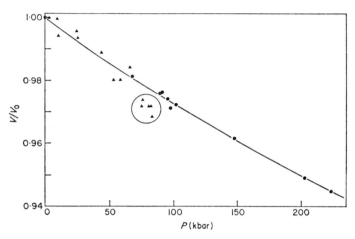

Fig. 36. Isothermal compression of stishovite at 25°C calculated from lattice parameters determined by X-ray diffraction. The triangles indicate data points obtained with the tetrahedral press (Bassett and Barnett, 1970); the circles indicate data points obtained with the diamond press (Liu *et al*, in preparation). The circled cluster of points are believed to have been influenced by deformation of the tetrahedral press.

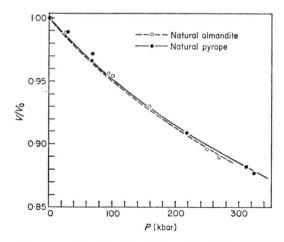

Fig. 37. Isothermal compression of natural almandite, $(Mg_{0.2}Fe_{0.8})_3Al_2Si_3O_{12}$, and natural pyrope, $(Mg_{0.6}Fe_{0.3})_3Al_2Si_3O_{12}$, at 25°C calculated from lattice parameters determined by X-ray diffraction (Takahashi and Liu, 1970).

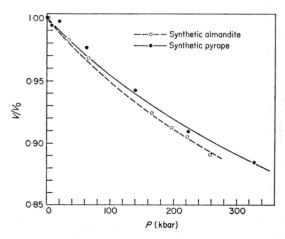

Fig. 38. Isothermal compression of synthetic almandite, $Fe_3Al_2Si_3O_{12}$, and synthetic pyrope, $Mg_3Al_2Si_3O_{12}$, at 25°C calculated from lattice parameters determined by X-ray diffraction (Takahashi and Liu, 1970).

(1970) and Liu *et al.* (in preparation) and the values are listed in Table 12. In comparing the data points from the two stishovite studies it should be noted that the first nine points obtained by the tetrahedral press are in good agreement with the data obtained by the diamond press and only the last five points are in real disagreement. They cluster below the curve fitted to the diamond press data. This cluster may be the result of deformation of the tetrahedral press once the sample gaskets have essentially ceased to be compressible. Such a deformation may introduce a significant error in the measurements of so incompressible a material as stishovite but not samples as compressible as NaCl. We conclude that the bulk modulus based on the data of Liu *et al.* is the more accurate.

b. *Bulk modulus systematics*   Anderson and Nafe (1965) pointed out that the bulk moduli of various oxides and silicates (such as stishovite, corundum, periclase, spinel, olivine, tourmaline and quartz) vary inversely as the fourth power of the volume per ion pair. They also found that the bulk moduli for alkaki halides and for base metal chalcogenides (such as ZnS, CdS, ZnSe, CdSe, ZnTe, and CdTe) are inversely proportional to the volume per ion pair. Anderson and Soga (1967) further extended the study to find that the bulk moduli for simple oxides such as BeO, MgO, ZnO and CaO are inversely proportional to the volume. They suggested that the crystals composed of ions having the same valencies should follow the inverse first power relation, whereas the crystalline species which have the same mean atomic weight but consist of ions having different valencies should follow the inverse fourth power relation.

Following Born and Huang (1954) and Anderson and Nafe (1965), we assume the lattice energy u to be in the form

$$u = -A/r + Be^{-r/q} \tag{1}$$

$$A = e^2 r \sum_{\substack{i=1 \\ i \neq j}}^{n} \sum_{j=1}^{\infty} Z_i Z_j / r_{ij} = \bar{a} e^2 (\bar{Z}_+ \bar{Z}_-) \tag{2}$$

$$B = M\lambda_{+-},$$

where r is the nearest interatomic distance; e is the electron charge; Z_i and Z_j are ionic valence numbers for ions i and j; $\bar{Z}_+$ and $\bar{Z}_-$ are the mean cation and anion valence numbers weighted according to $1/r_{ij}$; n is the number of ions in the molecular formula; r_{ij} is the interatomic distance between ions i and j; q and λ_{+-} are the constants related to the overlap potential between the cations and anions; $\bar{a}$ is the Madelung constant.

Since $P = -du/dV$, $V = cr^3$ (where c is a constant relating the volume to the nearest interatomic distance) and $K = -V(\partial P/\partial V)_T$, substitution of equation (1) into these equations yields

$$K_0 = \frac{A[-2 + (r_0/q)]}{9cr_0 V_0} \tag{3}$$

in which the boundary condition, $P = 0$ at $r = r_0$, was used to eliminate $Be^{-r_0/q}$.

For complex crystals such as spinels and garnets which have more than one cation species or position, the lattice energy and the bulk modulus are also a function of the cation species and positions as well as the atomic and lattice parameters mentioned above. However, since the lattice energy is an algebraic sum of bond energies, the bulk modulus will be given by a functional form similar to (3) in which each of the parameters represents a weighted mean of varying interatomic interactions (Sammis, 1970).

Further differentiation of equation (3) yields (Weaver, 1971b)

$$K_0' = [12 - 3(r_0/q) - (r_0/q)^2]/[6 - 3(r_0/q)].$$

Anderson et al. (1968) and others have reviewed the experimental data for K_0' and concluded that the K_0' values for nearly all of the oxides and silicates lie in the range between 4 and 7. This range of K_0' values corresponds to the (r_0/ρ) values of 7.4 to 16.1, which are all significantly greater than 2. The experimental data also indicate that the K_0' values for isostructural crystals consisting of the same anionic species differ by no more than 30 per cent, whereas the value for r_0 in the same crystal group does not vary more than 10 per cent. If the K_0' values for the isostructural crystals are assumed to be constant, then (r_0/q) should be constant as well. In this case, equation (3) is reduced to

$$K_0 \propto V^{-4/3}.$$

Another approach to simplifying equation (3) is to assume a constant value for q. Since q is a parameter related to repulsive force arising from electron overlap, it may not vary significantly for the isostructural crystals consisting of the same anionic species. Particularly in oxides and silicates, the repulsive forces are due mainly to anion–anion interactions. Therefore, this assumption may be well justified. In this case equation (3) may be reduced to

$$K_0 \approx (A/9cq)\, V_0^{-1}. \tag{4}$$

Since the value of A is a function of crystal structure and the valence of ions and since the value of c is related only to the crystal structure, the term $(A/9cq)$ is constant for isostructural crystals having the same valence. Thus, K_0 is inversely proportional to V_0. This relationship is the same as the inverse first power law of Anderson and Nafe (1965).

In studies of the Earth's interior, the ratio of the adiabatic bulk modulus to the density, (K_s/ρ), is conveniently used as it is directly related to the velocities of the compressional (v_p) and shear (v_s) elastic waves by

$$v_p^2 - \tfrac{4}{3}v_s^2 = K_s/\rho = \phi. \tag{5}$$

This quantity, which is commonly designated in the geophysical literature as ϕ or $(\partial P/\partial \rho)_s$, is called the seismic parameter. Its square root, $(\partial P/\partial \rho)_s^{1/2}$, is called hydrodynamic velocity. When q is assumed to be constant, equations (4) and (5) yield the relationship

$$\phi_0 = K_{s0}/\rho_0 \approx (A/9qcn)\, (1/\overline{M}), \tag{6}$$

where $\overline{M}$ is the mean atomic weight (formula weight per n), and n is the number of atoms in the formula, and the subscript $_0$ designates the quantity at zero pressure. For isostructural crystals, the term in the parentheses is nearly constant, and therefore ϕ_0 should be inversely proportional to the mean atomic weight. If K_0', hence (r_0/q), is assumed to be constant, we obtain

$$\phi_0 = \left(\frac{A}{9c^{4/3}n}\right) (r_0/q) (1/r_0\overline{M}).$$

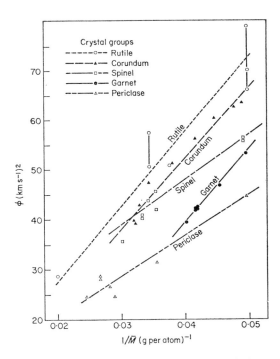

Fig. 39. Seismic parameter (ϕ) as a function of the reciprocal of mean atomic weight ($1/\overline{M}$) for five groups of oxides.

Figure 39 shows the ϕ versus $1/\overline{M}$ plot for the five isostructural crystal groups, and a linear relationship between ϕ and $1/\overline{M}$ can be observed. The values of ϕ and $1/\overline{M}$ for various crystalline species in the periclase $(M^{+2}O)$, corundum $(M_2^{+3}O_3)$, rutile $(M^{+4}O_2)$, spinel $(M^{+2}M_2^{+3}O_4$ or $M_2^{+2}M^{+4}O_4)$ and garnet $(M_3^{+2}M_2^{+3}M_3^{+4}O_{12})$ groups are listed in Tables 13 to 17. When the values of the adiabatic bulk modulus are not available, those values were computed from the isothermal values by the relation

$$K_s = K_t(1 + \alpha\gamma T),$$

TABLE 13

Mean atomic weight, bulk modulus, density and seismic parameter data for various periclase group crystals

Chemical composition	Density ρ_0 g cm^{-3}	Adiabatic bulk modulus K_{s0} Mbar	Mean atomic weight		ϕ_0 (Km s^{-1})^2	References
			$\bar{M}$ g per atom	$1/\bar{M}$ (g per atom)$^{-1}$ ×10^{-2}		
MgO Periclase	3.584	1.624	20.16	4.96	44.62	Schreiber and Anderson (1968)
CaO Lime	3.345	1.059	28.04	3.56	31.38	Soga (1968b)
MnO Manganosite	5.365	1.42[a]	35.47	2.82	26.5	Clendenen and Drickamer (1966)
CoO	6.438	1.85[a]	37.47	2.67	28.7	Clendenen and Drickamer (1966)
Fe$_{0.924}$O Wustite	5.745	1.44	34.61	2.89	24.7	Mao et al. (1969c)
NiO Bunsenite	6.808	1.95[a]	37.36	2.68	28.3	Clendenen and Drickamer (1966)

[a] These values were re-calculated from the data of Clendenen and Drickamer using the Birch–Murnaghan equation and the NaCl pressure scale of Weaver et al. (1971).

TABLE 14

Mean atomic weight and seismic parameter ϕ for various rutile group crystals

Crystalline species	Density ρ_0 g cm^{-3}	Mean atomic weight		Bulk modulus[a] at $P=0$ kbar		ϕ_0 (km s^{-1})^2	References
		$\overline{M}$ g per atom	$1/\overline{M}$ atoms per g $\times 10^{-2}$	K_t Mbar	K_s Mbar		
SiO$_2$ Stishovite	4.287	20.03	4.99	3.37 ± 0.13	3.39 ± 0.43	79.1 ± 3.0	Liu et al. (in preparation)
							Ahrens et al. (1970)
				3.0	3.0	70.0	Bassett and Barnett (1970)
				2.84 ± 0.25	2.86 ± 0.25	66 ± 6	
TiO$_2$ Rutile	4.250	26.63	3.76	2.136	2.155	50.72	Manghnani (1969)
MnO$_2$ Pyrolusite	5.233	28.98	3.45	2.6	2.6	50.6	McQueen and Marsh (1966)
							Weaver (personal communication)
SnO$_2$ Cassiterite	6.992	50.23	1.99	3.0	3.0	57.4	Ahrens et al. (1969)
				2.0	2.0	28.6	Madelung and Fuchs (1921)

[a] For the conversion of K_t to K_s, the relationship, $K_s = K_t(1+\alpha\gamma T)$ was used. For stishovite, $\alpha = (15\pm2) \times 10^{-6}$ K^{-1} and $\gamma = 1.58$; for rutile, $\alpha = 23.25 \times 10^{-1}$ K^{-1} and $\gamma = 1.28$ were used.

TABLE 15

Mean atomic weight and seismic parameter ϕ for various spinel group crystals

Chemical composition	Mineral or phase description	Mean atomic weight		$\phi_0{}^a$ (km s^{-1})^2	References
		$\bar{M}$ g per atom	$1/\bar{M}$ atom per g $\times 10^{-2}$		
Fe$_3$O$_4$	magnetite	33.08	3.023	35.6 ± 1.2	Mao (1967)
γ-Fe$_2$SiO$_4$	spinel phase	29.11	3.438	43.9 ± 2.0	Mao et al. (1969c)
γ-(Fe$_{0.9}$Mg$_{0.1}$)$_2$SiO$_4$	spinel phase	28.21	3.545	42.0 ± 2.0	Mao et al. (1969c)
γ-(Fe$_{0.8}$Mg$_{0.1}$)$_2$SiO$_4$	spinel phase	28.31	3.533	45.7 ± 0.2	Mao et al. (1969c)
γ-Co$_2$SiO$_4$	spinel phase	29.99	3.333	41.0 ± 1.2	Liu et al. (in preparation)
γ-Ni$_2$SiO$_4$	spinel phase	29.93	3.340	40.4 ± 1.4	Mao et al. (1970)
MgO·2.6Al$_2$O$_3$	spinel	20.36	4.912	55.82 ± 0.10	Schreiber (1967)
MgO·3.5Al$_2$O$_3$	spinel	20.37	4.910	55.87 ± 0.10	Verma (1960)

[a] The values for $\phi_0 (= K_s/\rho)$ were calculated from the adiabatic bulk modulus and density values for 25°C and 1 bar. The values of the adiabatic bulk modulus were computed from the isothermal bulk modulus values by an equation:

$$K_s = K_t(1 + \alpha \gamma T),$$

where α is the volume thermal expansion and γ is the Grüneisen parameter.

TABLE 16

Mean atomic weight, adiabatic bulk modulus, density and seismic parameter for various garnet group crystals

	Mean atomic weight		Density g cm^{-3}	K_s Mbar	ϕ_0 (km s^{-1})^2	References
	$\bar{M}$ g per atom	$1/\bar{M}$ atom per g $\times 10^{-2}$				
Mg$_3$Al$_2$Si$_3$O$_{12}$ Pyrope	20.16	4.96	3.564	1.91 ± 0.06	53.6	Takahashi and Liu (1970)
Py$_{60}$Alm$_{31}$[a]	21.93	4.56	3.81	1.78 ± 0.06	46.7	Takahashi and Liu (1970)
Py$_{22}$Alm$_{72}$[a]	23.79	4.20	4.16	1.770	42.52	Soga (1967)
	23.79	4.20	4.16	1.74 ± 0.06	41.8	Takahashi and Liu (1970)
Py$_{15}$Alm$_{78}$[a]	23.9	4.18	4.163	1.76	42.3	Adams and Gibson (1929)
Py$_{14}$Alm$_{81}$[a]	23.9	4.18	4.183	1.76	42.1	Verma (1960)
Fe$_3$Al$_2$Si$_3$O$_{12}$ Almandite	24.89	4.02	4.312	1.69 ± 0.05	39.2	Takahashi and Liu (1970)

[a] Py, pyrope; Alm, almandite. These are natural minerals and contain up to 3 mol per cent of grossularite and spessarite.

TABLE 17

Mean atomic weight and seismic parameter ϕ for various corundum group
crystals

Crystalline species	Mean atomic weight		ϕ_0 (km s^{-1})^2	References
	$\overline{M}$ g per atom	$1/\overline{M}$ atom per g $\times 10^{-2}$		
α–Fe$_2$O$_3$ Haematite	31.94	3.23	39.21	Liebermann and Schreiber (1968)
	31.94	3.23		
(MgFe)TiO$_3$ Ilmenite	31.28	3.20	39.9	Liu et al. (in preparation)
α–Al$_2$O$_3$ Corundum	20.39	4.90	63.22	Schreiber and Anderson (1966)
(Al$_{95}$Cr$_5$)$_2$O$_3$	20.89	4.79	62.5	Rossi (1966)
(Al$_{85}$Cr$_{15}$)$_2$O$_3$	21.89	4.56	60.4	Rossi (1966)
(Al$_{64}$Cr$_{36}$)$_2$O$_3$	24.00	4.17	56.1	Rossi (1966)
(Al$_{41}$Cr$_{59}$)$_2$O$_3$	26.30	3.80	51.2	Rossi (1966)
(Al$_{21}$Cr$_{79}$)$_2$O$_3$	28.30	3.53	47.1	Rossi (1966)
Cr$_2$O$_3$	30.40	3.29	42.8	Rossi (1966)

where α is the volume thermal expansion, γ is the Grünseisen parameter, and T is the temperature in K. Although the plot of ϕ versus $1/\overline{M}$ is a restatement of the inverse first-power law of Anderson and Nafe (1965), this has the advantage of representing the effect of chemical composition or $\overline{M}$ on the bulk modulus more clearly. The ionic valences for the various crystalline species are given in Table 18.

In Fig. 39, it is seen that the data points for the garnet group fall on a straight line better than those for any other group. Since the garnet data are mainly for ferromagnesian solid solutions and the K_0' values are about 5, the assumption that q is constant appears to be well justified for this case. On the other hand, the data for the spinel group include a wide variety of chemical compositions such as magnetite, silicates and Mg-Al spinels. Hence, the values for q in these may differ significantly. Similarly, the rutile and corundum groups include a wide variety of cation species and show wider scattering of data points than the garnet group. Thus, the linear relationship between ϕ and $1/\overline{M}$ appears to be best justified for solid solutions such as the garnet group.

TABLE 18

The ionic valences for various crystal groups

Crystal groups	General formulae	$\overline{Z}_+$	$\overline{Z}_-$	$(\overline{Z}_+\overline{Z}_-)$
Rutile	$M^{+4}O_2$	4	2	8
Corundum	$M_2^{+3}O_2$	23	32	6
Periclase	$M^{+2}O$	2	2	4
Spinel	$M^{+2}M_2^{+3}O_4$	8/3	2	5.3
	$M_2^{+2}M^{+4}O_4$	8/3	2	5.3
Garnet	$M_3^{+2}M_2^{+3}M_3^{+4}O_{12}$	24/8	2	6

3.3.2 *Phase transformation in magnetite*

Magnetite (Fe_3O_4) undergoes a reversible phase transformation at a pressure between 50 and 250 kbar. The structure of the high-pressure phase is as yet unidentified.

Mao *et al.* (1969a) have studied the effects of pressure on a natural sample of magnetite (Fe_3O_4) from Mineville, New York, by means of a diamond press modified to heat a sample to temperatures up to 310°C while it is under pressure. At room temperature and atmospheric pressure magnetite has a spinel structure with ferrous ions in six-fold coordination. The unit cell dimension of the starting sample is 8.394 ± 0.002 Å. In their investigation of the compression of magnetite at room temperature, they found that some new lines appeared in the diffraction pattern at approximately 250 kbar, indicating a partial transformation to a new phase. In an effort to drive the transformation to completion, they heated the sample to 310°C for a period of 30 minutes while it was under a pressure in excess of 250 kbar, and then obtained a diffraction pattern of it at room temperature while it was still under pressure. The *d*-spacings and intensities that they observed are given in Table 19. The decrease of the intensity of the (311) line of magnetite from $I/I_{10} = 10$ to $I/I_{10} = 2$, accompanied by an increase of the intensity of a completely new line $(d = 2.59$ Å$)$ to $I/I_{10} = 10$, indicates that although the transformation may not have gone to completion, it certainly has progressed most of the way.

Both the heated and unheated samples reverted to the spinel structure

TABLE 19

X-ray diffraction data for the high-pressure phase of Fe_3O_4

I/I_{10}	$d(\text{Å})$
10	2.59
2	2.44
2	2.33
0.5	2.13
3	2.02
3	1.89
1	1.79
3	1.55
5	1.40
1	1.23

having the original unit cell dimension after release of pressure. The reversion to the low-pressure phase did not start until pressures were below 50 kbar. Thus the equilibrium transformation pressure probably lies between 50 and 250 kbar. At atmospheric pressure some residual high-pressure phase remained for a period of a few weeks.

The structure of the high-pressure phase has not yet been determined. It cannot be indexed as cubic, tetragonal, or hexagonal with a reasonably small unit cell. It is very unlikely that it is a phase assemblage consisting of more than one phase as diffusion rates at room temperature would be insufficient for the recombination to form the spinel phase. Also, the pattern is not consistent with any of the phases that might be derived from magnetite, such as wustite or the hcp phase of iron. Although four of the lines (1, 2, 4, and 7) are remarkably similar to haematite lines at that pressure, haematite is stable at atmospheric oxygen pressure and would not revert to magnetite upon having its pressure released and being exposed to air.

Mao *et al.* (1969a) conclude that the new structure may be the result of the ferric iron ions being promoted from four-fold coordination sites to six-fold coordination sites.

3.3.3 *Pressure dependence of axial ratio in stishovite and ilmenite*

The two non-cubic oxides for which we have investigated the effect of pressure on the lattice parameters have each shown a pronounced pressure dependence of the axial ratio.

a. *Stishovite* There have been three X-ray diffraction studies on the effect of pressure on the lattice parameters of stishovite which has the tetragonal rutile structure. Ida *et al.* (1967) report that with increasing pressure the *c*-axis dimension first increases up to 90 kbar and then decreases up to 125 kbar, while the *a*-axis decreases at a steady rate. Bassett and Barnett (1970) report that both axes decrease with pressure and that the relative rates of decrease up to 80 kbar can be expressed by

$$\Delta c/c_0 = (0.4 \pm 0.2)\ \Delta a/a_0.$$

Liu *et al.* (in preparation), using the diamond anvil press, measured the effect of pressure on axial ratio up to 223 kbar and report the relationship

$$\Delta c/c_0 = (0.65 \pm 0.1)\ \Delta a/a_0 \quad \text{(Figs 40 and 41)}.$$

We feel this third determination is probably the most accurate since the measurements were made over a larger pressure range than that used in the other two studies. Also, the lattice parameter determinations show less scatter than those of Bassett and Barnett (1970).

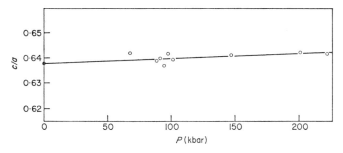

Fig. 40. The axial ratio (*c*/*a*) of stishovite, which is tetragonal, as a function of pressure at 25°C determined by X-ray diffraction (Liu *et al.*, in preparation).

The anisotropy of the compressibility of stishovite can be treated in terms of the compressibilities of the individual ion pairs and their orientations in the structure. A treatment of this type is consistent with the observations reported above.

The difference in linear compressibilities parallel to the *c*-axis and perpendicular to the *c*-axis must reflect the anisotropy of the atomic configuration and bonding in stishovite, which has the rutile structure. In this structure it is possible to find straight lines or chains of nearest neighbour ions. Parallel to the *c*-axis there are chains consisting of

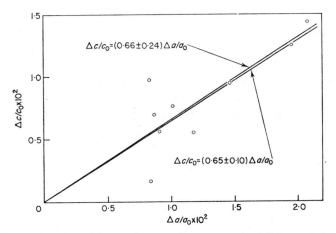

Fig. 41. The compression of the *a*-axis versus the compression of the *c*-axis in stishovite. This relationship gives a more sensitive indication of the change in axial ratio than the previous diagram (Liu *et al.*, in preparation). The upper curve was fitted to all of the data points; the lower curve was fitted to the data points excluding the two that deviate the most.

–oxygen–oxygen–oxygen– while parallel to the [110] axis there are chains consisting of –silicon–oxygen–oxygen–silicon–oxygen–oxygen–. There are nearly twice (1.81) as many chains parallel to the *c*-axis as parallel to the [110] axis. Furthermore, the oxygen–oxygen interionic distances parallel to the *c*-axis are short compared with the sums of the ionic radii, indicating rather extensive overlap of electron orbitals and low compressibility. The silicon–oxygen distances in the chains parallel to the [110] axis are nearly equal to the sums of the ionic radii, indicating little overlap of electron orbitals and a higher compressibility. Thus, both the numbers of chains and the spacings of the ions in the chains are consistent with the observed difference in linear compressibilities of stishovite.

b. *Ilmenite* Liu *et al.* (in preparation) have measured the effect of pressure on the lattice parameters of ilmenite (hexagonal) up to 275 kbar and have observed that the compression parallel to the *c*-axis is greater than the compression parallel to the *a*-axis. They have expressed the relationship by the equation

$$\Delta c/c_0 = (1.73 \pm 0.15)\, \Delta a/a_0 \qquad \text{(Fig. 42).}$$

As in stishovite, the difference in linear compressibility parallel to and perpendicular to the *c*-axis can be explained in terms of ion pairs

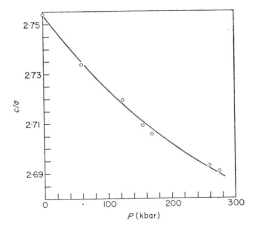

Fig. 42. The axial ratio (c/a) of ilmentite, which is hexagonal, as a function of pressure at 25°C determined by X-ray diffraction (Liu *et al.*, in preparation).

and their orientations in the structure. In this structure the iron ion and the titanium ion occupy positions just above and just below each triangular group of oxygen ions. They are on the trigonal axis passing through the oxygen group. Their distances from the oxygens in the triangular group are rather large compared with the sums of ionic radii. If these distances decrease more rapidly with compression than the other ionic distances, a greater compressibility along the c-axis than perpendicular to it should result.

Rotation of the triangular groups of oxygen is permitted by the space group $(R\bar{3})$ of ilmenite. Such a motion would have the effect of creating space along the c-axis to accommodate the larger iron ions, thus causing the c-axis to be more compressible than the directions perpendicular to the c-axis.

4 Fixed point calibrations above 100 kbar

The iron α–ϵ transition and the lead I–II transition have been used as calibration points on the fixed point scale because of the discontinuous change in electrical resistance at the transitions. Bancroft *et al.* (1956), using shock loading, reported a phase transformation in pure iron at 130 kbar. Balchan and Drickamer (1961), basing their determinations on electrical resistance made in a Drickamer-type cell, reported that the transformation in iron takes place at 133 kbar, and that in lead at

161 kbar. For several years these values for the iron and lead transitions have been used for calibration.

When we attempted to use these transitions for calibration of a Drickamer cell, we found that they both occurred at very nearly the same load. We then examined a mixture of iron and lead in the diamond anvil press by X-ray diffraction (Mao *et al.*, 1969b) and found that the most intense lines of the high-pressure phase of iron began to appear at a pressure just below the pressure at which the line of the high-pressure phase of lead began to appear. Using the NaCl pressure scale of Weaver *et al.* (1971) we determined that the iron transition was occurring between 107 and 127 kbar and the lead transition was occurring between 128 and 131 kbar (Table 20).

TABLE 20

Transition pressures for iron and lead

Investigators	Iron kbar	Lead kbar
Mao *et al.* (1969b)	107–123	128–131
Vereshchagin *et al.* (1970)	127–129	135–138
Drickamer (1971)	110–113	128–132

Drickamer (1971), in his revised fixed-point pressure scale, now places the pressure of the iron transition at 110–113 kbar and the lead transition at 128–132 kbar. Vereshchagin *et al.* (1970) also published a revised fixed-point scale in which they place the iron transition at 127–129 kbar and the lead transition at 135–138 kbar. Thus, there seems to be fair agreement among the three groups concerning the pressures of the transitions in iron and lead.

In recent years high-pressure X-ray diffraction has provided a means for measuring sample pressure that is independent of load. The results of investigations of phase transitions used as points on the fixed-point pressure scale have in general resulted in downward revisions (Jeffery *et al.*, 1966; Mao *et al.*, 1969b; Vereshchagin *et al.*, 1970; Drickamer, 1971). This probably indicates that earlier extrapolations of pressure-load relationships failed to consider the rapidity with which the efficiency of high-pressure devices falls off with increasing pressure. The curve shown in Fig. 4 illustrates this for the diamond anvil cell. Figure 1 in Drickamer (1971) shows a very similar relationship for the Drickamer cell.

In conclusion, we would like to urge that caution be used in the extrapolation of pressure scales on the basis of load.

5 Conclusions

1. The results reported in this review were obtained by means of X-ray diffraction measurements of samples while the samples were under pressure in a diamond anvil press. Molar volumes were calculated from the lattice parameters of the samples and pressures were calculated from the lattice parameters of NaCl or CsI intimately mixed with the sample. A vibrational formulation of the Hildebrand equation has provided a pressure–volume relationship for NaCl that is in good agreement with other theoretical models as well as experimental measurements. The semi-empirical Fritz–Thurston equation (1970) combined with the ultrasonic elastic parameters, K_0, K_0', and K'', obtained by Barsch and Chang (1971) for CsI yields a P–V relation consistent with the results of X-ray diffraction measurements. Either NaCl or CsI may be used as a pressure standard.

2. The volume changes accompanying the B1–B2 phase transformations have been measured in NaCl, KCl, KBr, and KI. Entropy changes have been calculated from these volume data and from the slopes of the phase boundaries by means of the Clapeyron equation. We have formulated a thermodynamic equation which relates the entropy changes and the volume changes for the alkali halides. The four alkali halides which we have studied plus four others for which information is available are consistent with this relationship, and yield a value of 2.3 eu for the entropy increase that would take place if an alkali halide were to undergo a B1–B2 phase transformation with zero volume change. We have called this the configurational entropy change.

3. High-pressure X-ray diffraction studies by us and by others have shown that the B1 phases of silver halides undergo pressure-induced phase transformations to phases having lower symmetry than the B2 phase to which the alkali halides transform. This behaviour can probably be attributed to the more covalent and hence more directional nature of the bonding in the silver halides.

4. Compression measurements have been made on iron and its nickel

and silicon alloys, on lead, and on rhenium. It was found that the compressibility of the bcc phases of the iron-nickel alloys increases linearly with increasing nickel, while the compressibility of the bcc phases of the iron-silicon alloys decreases linearly with increasing silicon. The volume change accompanying the bcc–hcp phase transformation in the iron-nickel alloys increases with increasing nickel. whereas it decreases with increasing silicon in the iron-silicon alloys. Determinations of the transformation pressures for these alloys are consistent with the shock determinations which indicate that the transformation pressure decreases with increasing nickel and increases with increasing silicon. The Fe-25Si, which forms an intermetallic compound, does not undergo a phase transformation from the bcc to the hcp structure up to 238 kbar. This result is consistent with that of the shock studies of Fe-31Si of Balchan Cowan (1966) who failed to detect a phase transformation up to 1000 kbar.

5. The phase transformation in lead has a volume change of only -0.18 cm^3 mol^{-1}. This is apparently so small a change that the shock method is unable to detect it. X-ray diffraction measurements of the pressure of the phase transformation in lead using NaCl pressure standard yield a pressure lying in the range between 128 and 131 kbar, a value which is considerably lower than earlier published values.

6. The hcp phases of metals which we have studied show no variation of c/a ratio with pressure. The ratios lie in the range 1.603 to 1.650. We conclude that the ratios remain constant because they are so close to the ideal axial ratio of 1.633 for the packing of hard spheres. The axial ratio of rhenium was measured as a function of temperature and found to decrease with increasing temperature. We attributed this to a difference in anharmonicity in different directions in the structure.

7. Compression measurements have been made on several oxides and silicates. From these measurements the seismic parameters $\phi = (\partial P/\partial \rho)_s$ have been calculated and plotted against the reciprocal of the mean atomic weight, $1/\overline{M}$. The data points for the ferromagnesian garnet group fall on a straight line, while those for other groups having a wider variety of cationic compositions show a greater scatter. Thus, the linear relationship seems to be best justified for simple solid solutions such as the ferromagnesian garnet series.

8. Laue photographs of the diamond anvils under load indicate that the anvils elastically deform to become cup-shaped with a radius of curvature of the order of millimetres. This deformation is partially

responsible for the loss in efficiency of the anvil apparatus (ratio of measured pressure to load). We suggest that a similar loss of efficiency takes place in other high-pressure devices and therefore caution should be used when extrapolating a pressure scale on the basis of a pressure-load relationship.

ACKNOWLEDGMENTS

We wish to express our profound thanks for the collaboration, assistance and stimulating discussions given to us by the former and present members of our research team at the University of Rochester: Ho-Kwang Mao, Lin-Gun Liu, J. Scott Weaver, Douglas E. Hammond, Miles L. Silberman, Phillip W. Stook and Josephine K. Campbell. We also thank Jane K. Bassett for preparing the illustrations. We are greatly indebted to the National Science Foundation for providing financial support for the research program.

References

Adams, L. H. and Gibson, R. E. (1929). *Nat. Acad. Sci.-Nat. Res. Counc. Publ.* **15**, 713.

Adams, L. H. and Davis, B. L. (1965). *Amer. J. Sci.* **263**, 359.

Ahrens, T. J., Anderson, D. L. and Ringwood, A. E. (1969). *Rev. Geophys.* **7**, 667.

Ahrens, T. J., Takahashi, T. and Davies, G. (1970). *Geophys. Res. Pap.* **75**, 310.

Alder, B. J. (1963). *In* "Solids Under Pressure" (Ed. W. Paul and D. M. Warschauer). McGraw-Hill, New York.

Alt'shuler, L. V., Kormer, S. V., Bakanova, A. A. and Trunin, R. F. (1960). *Zh. Eksp. Teor. Fiz.* **38**, 790.

Alt'shuler, L. V., Kuleshova, L. V., and Pavlovskii, M. N. (1961). *Sov. Phys.-JETP*, **12**, 10.

Anderson, O. L. (1967). *J. Geophys. Res.* **72**, 3661.

Anderson, O. L. and Andreatch, P. (1966). *J. Amer. Ceram. Soc.* **49**, 404.

Anderson, O. L. and Nafe, J. E. (1965). *J. Geophys. Res.* **70**, 3951.

Anderson, O. L. and Soga, N. (1967). *J. Geophys. Res.* **72**, 5754.

Anderson, O. L., Schreiber, E. and Lieberman, R. C. (1968). *Rev. Geophys.* **6**, 491.

Balchan, A. S. and Cowan, G. R. (1966). *J. Geophys. Res.* **71**, 3577.

Balchan, A. S. and Drickamer, H. G. (1961). *Rev. Sci. Instrum.* **32**, 308.

Bancroft, D., Peterson, E. L. and Minshall, F. S. (1956). *J. Appl. Phys.* **27**, 291.

Barnett, J. D. and Hall, H. T. (1964). *Rev. Sci. Instrum.* **35**, 175.

Barsch, G. R. and Chang, Z. P. (1971). *In* "Accurate Characterization of the High Pressure Environment" (Ed. E. Lloyd), p. 173. Special Publication, 326, National Bureau of Standards.

Bartels, R. A. and Schuele, D. E. (1965). *J. Phys. Chem. Solids*, **26**, 537.

Bassett, W. A. and Barnett, J. D. (1970). *Phys. Earth Planet, Interiors*, **3**, 54.

Bassett, W. A. and Takahashi, T. (1964). *ASME* 64-*WA/PT*-24, American Society of Mechanical Engineers.

Bassett, W. A. and Takahashi, T. (1965). *Amer. Mineral.* **50**, 1576.

Bassett, W. A., Takahashi, T. and Stook, P. W. (1967). *Rev. Sci. Instrum.* **38**, 37.

Bassett, W. A., Takahashi, T., Mao, H. K. and Weaver, J. S. (1968). *J. Appl. Phys.* **39**, 319.

Bassett, W. A., Takahashi, T. and Campbell, J. K. (1969). *Trans. Amer. Crystallogr. Ass.* **5**, 93.

Birch, F. (1947). *Phys. Rev.* **71**, 809.

Blackburn, L. D., Kaufman, L. and Cohen, M. (1965). *Acta Met.* **13**, 533.

Born, M. and Huang, K. (1954). "Dynamical Theory of Crystal Lattices". Oxford University Press, London.

Bradley, R. S., Grace, J. D. and Munro, D. C. (1964). *Z. Kristallogr. Mineral.* **120**, 349.

Bridgman, P. W. (1915). *Proc. Amer. Acad. Arts Sci.* **51**, 55.

Bridgman, P. W. (1940). *Proc. Amer. Acad. Arts. Sci.* **74**, 21.

Bridgman, P. W. (1945). *Proc. Amer. Acad. Arts Sci.* **76**, 1.

Bridgman, P. W. (1948). *Proc. Amer. Acad. Arts Sci.* **76**, 55.

Bridgman, P. W. (1949). "The Physics of High Pressure". G. Bell and Sons, London.

Bridgman, P. W. (1955). *Proc. Amer. Acad. Arts Sci.* **84**, 112.

Brugger, R. M., Bennion, R. B. and Worlton, T. G. (1967). *Phys. Lett.* **A24**, 714.

Brugger, R. M., Bennion, R. B., Worlton, T. G. and Myers, W. R. (1969). *Trans. Amer. Crystallogr. Ass.* **5**, 141.

Bundy, F. P. (1965). *J. Appl. Phys.* **36**, 616.

Burley, G. (1963). *Amer. Mineral.* **48**, 1266.

Burley, G. (1964). *J. Phys. Chem.* **68**, 1111.

Chang, Z. P. and Barsch, G. R. (1967). *Phys. Rev. Lett.* **19**, 1381.

Christian, R. H. (1957). Lawrence Radiation Laboratory Report UCRL-4900.

Clendenen, R. L. and Drickamer, H. G. (1964). *J. Phys. Chem. Solids,* **27**, 291.

Clendenen, R. L. and Drickamer, H. G. (1966). *J. Chem. Phys.* **44**, 4223.

Cohn, W. M. (1933). *Phys. Rev.* **44**, 326.

Darnell, A. J. and McCollum, W. A. (1970). *J. Phys. Chem. Solids,* **31**, 805.

Davis, B. L. and Adams, L. H. (1962). *Z. Kristallogr. Mineral.* **117**, 399.

Davis, B. L. and Adams, L. H. (1964). *Science,* **146**, 519.

Decker, D. L. (1965). *J. Appl. Phys.* **36**, 157.

Decker, D. L. (1966). *J. Appl. Phys.* **37**, 5012.

Decker, D. L., Bassett, W. A., Merrill, L., Hall, H. T. and Barnett, J. D. (In press). National Bureau of Standards Circular.

Drickamer, H. G. (1971). *Rev. Sci. Instrum.* **41**, 1667.

Drickamer, H. G., Lynch, R. W., Clendenen, R. L. and Perez-Albuerne, E. A. (1966). *Solid State Phys.* **19**, 135.

Farquhar, M. C. M., Lipson, H. and Weill, A. R. (1945). *J. Iron Steel Inst.* **152**, 457.

Fowler, C. M., Minshall, F. S. and Zukas, E. G. (1961). *In* "Responses of Metals to High Velocity Deformation" (Eds Shewnon and Zackay), p. 275. Interscience, New York.

Freud, P. J. and LaMori, P. N. (1969). *Trans. Amer. Crystal. Assoc.* **5**, 155.

Freud, P. J. and Sclar, C. B. (1969). *Rev. Sci. Instrum.* **40**, 434.

Frevel, L. K. (1935). *Rev. Sci. Instrum.* **6**, 214.

Fritz, J. N., Marsh, S. P., Carter, W. J. and McQueen, R. G. (1971). *In* "Accurate Characterization of the High Pressure Environment" (Ed. E. Lloyd), p. 201. Special Publication 326, National Bureau of Standards.

Fritz, T. C. and Thurston, R. N. (1970). *J. Geophys. Res.* **75**, 1557.

Fumi, F. G. and Tosi, M. P. (1962). *J. Phys. Chem. Solids*, **23**, 395.

Hajj, F. (1966). *J. Chem. Phys.* **44**, 4618.

Hammond, D. E. (1969). "The Pressure Volume Relationship of CsI up to 250 kb at Room Temperature". Thesis, MS, University of Rochester, Rochester, N.Y.

Haussühl, S. (1960). *Z. Phys.* **159**, 223.

Hauver, G. E. and Melani, A. (1970). *Ballistics Res. Lab. Mem. Rep.* **2061**, 1.

Haygarth, J. C., Getting, I. C. and Kennedy, G. C. (1967). *J. Appl. Phys.* **38**, 4557.

Haygarth, J. C., Luedmann, H. D., Getting, I. C. and Kennedy, G. C. (1969). *J. Phys. Chem. Solids*, **30**, 1417.

Ida, Y., Syono, Y. and Akimoto, S. (1967). *Earth Planet. Sci. Lett.* **3**, 216.

Jacobs, R. B. (1938). *Phys. Rev.* **54**, 325.

Jamieson, J. C. (1957). *J. Geol.* **65**, 334.

Jamieson, J. C. (1961). *In* "Progress in Very High Pressure Research" (Eds Bundy, Hibbard and Strong). John Wiley, New York.

Jamieson, J. C. and Lawson, A. W. (1962). *J. Appl. Phys.* **33**, 776.

Jeffery, R. N., Barnett, J. D., Vanfleet, H. B. and Hall, H. T. (1966). *J. Appl. Phys.* **37**, 3172.

Johnson, J. W., Agron, P. A. and Bredig, M. A. (1955). *J. Amer. Chem. Soc.* **77**, 2734.

Johnson, Q. (1966). *Science*, **153**, 419.

Kasper, J. S. (1960). *J. Metals*, **12**, 732.

Kaufman, L., Clougherty, E. V. and Weiss, R. J. (1963). *Acta Met.* **11**, 323.

Kennedy, G. C. and LaMori, P. N. (1961). *In* "Progress in Very High Pressure Research" (Eds Bundy, Hibbard and Strong). John Wiley, New York.

Klement, W., Jr. (1963). *J. Chem. Phys.* **38**, 298.

Kuzubov, A. (1965). Lawrence Radiation Laboratory Report UCRL-12473.

Lawson, A. W. and Riley, N. A. (1949). *Rev. Sci. Instrum.* **20**, 763.

Lawson, A. W. and Tang, T. Y. (1950). *Rev. Sci. Instrum.* **21**, 815.

Leibfried, G. and Ludwig, W. (1961). *In* "Solid State Physics" (Eds F. Seitz and D. Turnbull), Vol. 12, p. 275. Academic Press, New York and London.

Liebermann, R. and Schreiber, E. (1968). *J. Geophys. Res.* **73**, 6585.

Liu, L. G., Takahashi, T. and Bassett, W. A. (1970). *J. Phys. Chem. Solids*, **31**, 1345.

Madelung, E. and Fuchs, R. (1921). *Ann. Phys.* **65**, 20.

Majumdar, A. J. and Roy, R. (1959). *J. Phys. Chem.* **63**, 1858.

Manghnani, M. H. (1969). *J. Geophys. Res.* **74**, 4317.

Mao, H. K. (1967). The pressure dependence of lattice parameters and volume of ferromagnesian spinels, and its implications to the Earth's mantle. Ph.D. Thesis, University of Rochester, Rochester, New York.

Mao, H. K., Bassett, W. A. and Takahashi, T. (1967). *J. Appl. Phys.* **38**, 272.

Mao, H. K., Bassett, W. A. and Takahashi, T. (1969a). Annual Report of the Director, Geophysical Laboratory, Carnegie Institute Yearbook, Vol. 68, p. 249.

Mao, H. K., Takahashi, T. and Bassett, W. A. (1969b). Annual Report of the Director, Geophysical Laboratory, Carnegie Institute Yearbook, Vol. 68, p. 249.

Mao, H. K., Takahashi, T., Bassett, W. A., Weaver, J. S. and Akimoto, S. (1969c). *J. Geophys. Res.* **74**, 1061.

Mao, H. K., Takahashi, T. and Bassett, W. A. (1970). *Phys. Earth Planet. Interiors*, **3**, 51.

Mayer, J. E. (1933). *J. Chem. Phys.* **1**, 270.

McQueen, R. G. and Marsh, S. P. (1960). *J. Appl. Phys.* **31**, 1253.

McQueen, R. G. and Marsh, S. P. (1966). *J. Geophys. Res.* **71**, 1751.

McSkimin, H. J. (1950). *J. Acoust. Soc. Amer.* **22**, 413.

McWhan, D. B. (1965). *J. Appl. Phys.* **36**, 664.

McWhan, D. B. (1967). *J. Appl. Phys.* **38**, 347.

McWhan, D. B. and Bond, W. L. (1964). *Rev. Sci. Instrum.* **35**, 626.

Miller, R. A. and Schuele, D. E. (1969). *J. Phys. Chem. Solids*, **30**, 586.

Moore, M. J. and Kasper, J. S. (1968). *J. Chem. Phys.* **48**, 2446.

Murnaghan, F. D. (1937). *Amer. J. Math.* **49**, 235.

Neuhaus, A. and Hinze, E. (1966). *Ber. Bunsenges. Phys. Chem.* **70**, 1073.

Owen, N. B., Smith, P. L., Martin, J. E. and Wright, A. J. (1963). *J. Phys. Chem. Solids*, **24**, 1519.

Perez-Alburne, E. A. and Drickamer, H. G. (1965). *J. Chem. Phys.* **43**, 1381.

Perez-Albuerne, E. A., Forgren, K. F. and Drickamer, H. G. (1964). *Rev. Sci. Instrum.* **35**, 29.

Piermarini, G. J. and Weir, C. E. (1962). *J. Res. Nat. Bur. Stand.* **A66**, 325.

Pistorius, C. W. F. (1964). *J. Phys. Chem. Solids*, **25**, 1477.

Pistorius, C. W. F. (1965a). *J. Phys. Chem. Solids*, **26**, 1543.

Pistorius, C. W. F. (1965b). *J. Chem. Phys.* **43**, 1557.

Rice, M. H., McQueen, R. G. and Walsh, J. M. (1958). *In* "Solid State Physics" (Eds F. Seitz and D. Turnbull), Vol. 6, p. 1. Academic Press, New York and London.

Riggleman, B. M. and Drickamer, H. G. (1963). *J. Chem. Phys.* **38**, 2721.

Rossi, L. (1966). The elastic properties of oxide solids solutions; the system Al_2O_3–Cr_2O_3. Ph.D. Thesis, S.U.N.Y. College of Ceramics, Alfred University, New York.

Rotter, C. A. and Smith, C. S. (1966). *J. Phys. Chem. Solids*, **27**, 267.

Sammis, C. G. (1970). *Geophys. J. Roy. Astron. Soc.* **19**, 285.

Schock, R. N. and Jamieson, J. C. (1969). *J. Phys. Chem. Solids*, **30**, 1527.

Schreiber, E. (1967). *J. Appl. Phys.* **38**, 2508.

Schreiber, E. and Anderson, O. L. (1966). *J. Amer. Ceram. Soc.* **49**, 184.

Schreiber, E. and Anderson, O. L. (1968). *J. Geophys. Res.* **73**, 2837.

Shepard, M. L. and Smith, J. F. (1965). *J. Appl. Phys.* **36**, 1447.

Silberman, M. L. (1967). Isothermal compression of two iron-silicon alloys to 290 kilobars at 23°C. MS Thesis, University of Rochester, Rochester, New York.

Slagle, O. D. and McKinstry, H. A. (1967). *J. Appl. Phys.* **38**, 437.

Slykhouse, T. E. and Drickamer, H. G. (1958). *J. Phys. Chem. Solids*, **7**, 207.

Soga, N. (1967). *J. Geophys. Res.* **72**, 4227.

Soga, N. (1968a). *J. Geophys. Res.* **73**, 827.

Soga, N. (1968b). *J. Geophys. Res.* **73**, 5385.

Swanson, H. E. and Fuyat, R. K. (1953). *Nat. Bur. Stand. (US) Circ.* 539, **II**, 63.

Swanson, H. E., Fuyat, R. K., Ugrinic, G. M. (1953). *Nat. Bur. Stand. (US) Circ.* 539, **4**, 47.

Takahashi, T. and Bassett, W. A. (1964). *Science,* **145**, 483.

Takahashi, T., Bassett, W. A. and Mao, H. K. (1968). *J. Geophys. Res.* **73**, 4717.

Takahashi, T., Mao, H. K. and Bassett, W. A. (1969). *Science,* **165**, 1352.

Takahashi, T. and Liu, L. G. (1970). *J. Geophys. Res.* **75**, 5757.

Thomsen, L. (1970). *J. Phys. Chem. Solids,* **31**, 2329.

Tosi, M. P. (1964). *In* "Solid State Physics" (Eds F. Seitz and D. Turnbull). Vol. 16. Academic Press, New York and London.

Vaidya, S. N. and Kennedy, G. C. (1970). *J. Phys. Chem. Solids,* **31**, 2329.

Vaidya, S. N. and Kennedy, G. C. (1971). *J. Phys. Chem. Solids,* **32**, 951.

Van Valkenburg, A. (1964). *J. Res. NBS A68*, 97.

Vereshchagin, L. F. (1965). *In* "Physics of Solids at High Pressures" (Eds Tomizuka and Emrick). Academic Press, New York and London.

Vereshchagin, L. F., Semerchan, A. A., Kuzin, N. N. and Sadkov, Yu. A. (1969). *Sov. Phys.-Dokl.* **14**, 340.

Vereshchagin, L. F., Semerchan, A. A., Kuzin, N. N. and Sadkov, Y. A. (1970). *Sov. Phys.-Dokl.* **15**, 292.

Verma, R. K. (1960). *J. Geophys. Res.* **65**, 757.

Wagner, G. and Lippert, L. (1936). *Z. Phys. Chem.* **31**, 263.

Weaver, J. S., Takahashi, T. and Bassett, W. A. (1971). *In* "Accurate Characterization of the High Pressure Environment" (Ed. E. Lloyd), p. 189. Special Publication 326, National Bureau of Standards.

Weaver, J. S. (1971a). *In* "Accurate Characterization of the High Pressure Environment" (Ed. E. Lloyd), p. 325. Special Publication 326, National Bureau of Standards.

Weaver, J. S. (1971b). Equation of State of NaCl, MgO, and Stishovite. Ph.D. Thesis, University of Rochester, Rochester, New York.

Weir, C. E. and Piermarini, G. J. (1964). *J. Res. NBS 68A*, 105.

Zukas, E. G., Fowler, C. M., Minshall, F. S. and O'Rourk, J. (1963). *Trans. AIMME,* **227**, 746.

Diamond Formation at High Pressures

R. H. WENTORF, Jr

General Electric Company Research and Development Center, Schenectady, New York, USA

1	Introduction	249
2	Phase diagram for pure carbon	250
3	Relatively pure carbon systems	252
	3.1 Crystal structures	252
	3.2 Hexagonal diamond formation	253
	3.3 Cubic diamond formation	254
	3.4 Pyrolysis of carbonaceous materials	257
4	Carbon in solution	257
	4.1 Types of solvents	257
	4.2 Graphite-producing solvents	258
	4.3 Diamond-producing solvents	259
5	Growth of large laboratory diamond cyrstals	267
6	Impurities in diamonds	270
7	Polycrystalline diamond masses	274
8	Unexplored territory	279
	References	279

Abstract

Variety in the behaviour of carbon is not confined to low pressures, but exists at high pressures also. Diamond may form from a number of materials via many pathways. Despite the great accumulation of laboratory knowledge, the formation processes for most natural diamonds remain unknown.

1 Introduction

Carbon in its pure or nearly pure state may take a variety of forms, ranging from essentially amorphous chars and soots to well-crystallized graphite or diamond. The strength and directionality of the chemical bonds between carbon atoms give carbon its refractory nature and account for the comparative stability of the various forms of solid carbon. This stability is so great that frequently extremely high temperatures or strong solvents are required to disrupt a carbonaceous structure which has been built up via a particular chemical route or from a particular starting material.

This diverse and refractory nature of carbon, so well observed at low pressures where graphite is the ultimate crystalline product, might also be expected to manifest itself at high pressures where diamond is the ultimate crystalline form, although one would not expect that the high-pressure behaviour would be a "carbon copy" of that at low pressures.

In recent years the advancing techniques of high-pressure experimentation have made it possible to study some of the behaviour of carbon at high pressures. The purpose of this article is to gather together many of the results of such high-pressure experimentation on carbon which is connected with the formation of diamond. The many remaining mysteries about the formation of natural diamonds are sufficient to ensure that this body of high-pressure work is still incomplete; when dealing with an element as profound as carbon, omniscience is a distant goal, but the way to that goal can be fascinating.

Several workers (Eversole, 1961; Angus et al., 1968; Deryagin et al., 1970) have demonstrated the growth of diamond from the pyrolysis of methane on diamond seed crystals at low pressures. Some graphite also forms. Such processes are extremely interesting but are somewhat out of bounds for this article and so are mentioned in passing.

Similarly, the many developments in the art of high-pressure generation upon which the study of carbon at high pressures naturally depends are not described in this paper. The reader interested in these aspects will find descriptions of apparatus among the references mentioned in this article. An overall view of diamond synthesis, in great detail, may be found in a paper by Bundy, Strong and Wentorf (1972).

2   Phase diagram for pure carbon

The existing pressure and temperature exert a strong but not always overwhelming influence upon whether graphite or diamond will form in a given process which yields elemental carbon. Thermodynamic considerations tell us what the ultimate crystalline form will be, but "ultimate" implies that we may have to wait for an inconveniently long time.

The carbon phase diagram shown in Fig. 1 embodies the results of modern experimental work. As a start, we note that the density of perfect graphite at room temperature and pressure is about 2.25 g cm^{-3}

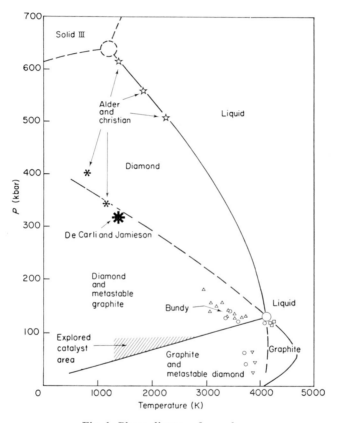

Fig. 1. Phase diagram for carbon.

and the volume of a gram-atom is about 5.33 cm^3; the density of diamond is considerably higher, about 3.52 g cm^{-3}, and the volume of a gram-atom is only 3.41 cm^3, some 64 per cent of that of graphite. Evidently high pressures will favour diamond. Up to about 1200 K the data on heat capacity, density, thermal expansion, compressibility and heats of combustion of graphite and diamond define the course of the equilibrium line tolerably well (Rossini and Jessup, 1938). For higher temperatures, about 2000 K, the experimental observation of diamond and graphite formation under near equilibrium conditions serve as a guide (Bundy et al., 1961). The considerations of Berman and Simon (1955) agree well with these observations and permit a modest extrapolation on the basis of $P = 7 + 0.027\ T$, where P is in kilobars and T is in K. For temperatures exceeding about 2000 K, the uncertainty of the position of the line increases owing to the difficulty of measuring accurately such high pressures and temperatures and the dearth of experimental data or guaranteed accurate models for calculation. Bundy's paper (1969) contains the results of work on the phase diagram of carbon under these most extreme conditions, including the melting of diamond and graphite. For pressures exceeding 200 kbar one must rely upon data from shock compression experiments. The work of Alder and Christian (1961) suggests that the melting point of diamond falls with increasing pressure, just as with Si and Ge.

3 Relatively pure carbon systems

3.1 CRYSTAL STRUCTURES

Figure 2 shows the crystal structures of perfect graphite, cubic diamond, and hexagonal diamond. Hexagonal diamond is related to ordinary cubic diamond as wurtzite is related to zinc blende; the carbon–carbon distances are essentially the same, but the layers of wrinkled hexagons are stacked in a different way. Hexagonal diamond was merely a postulated structure until 1967, when it was identified in several materials (Bundy and Kasper, 1967).

In graphite and hexagonal diamond the layer stacking sequence is

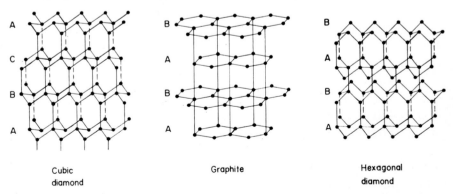

Cubic diamond   Graphite   Hexagonal diamond

Fig. 2. Crystal structures of cubic diamond, graphite, and hexagonal diamond.

ABAB . . . i.e. any A layer is simply a copy, vertically displaced along the c-axis, of the A layer beneath it. A B layer is an A layer which has been moved both vertically and horizontally; any B layer is a vertically displaced copy of the B layer beneath it. In cubic diamond the layer stacking sequence is ABCABC . . . ; again all the layers, if considered separately, would look the same, but they are stacked in a particular regular arrangement of groups of three to form cubic diamond.

With such identified layers in mind, one could visualize more complicated stacking sequences having longer repeat distances, such as ABCABABCAB . . . etc., to form other types of diamond, after the manner of the many polymorphs of SiC. Indeed such arrangements may actually exist, but for practical purposes of identification by X-ray diffraction, where a regular structure of say 100 or more repeating units is needed to produce satisfactory diffraction patterns, the two ideal types of diamond indicated in Fig. 2 suffice today to describe the crystalline products obtained in diamond-making experiments. It appears that cubic diamond is slightly more stable than hexagonal diamond, i.e. the latter forms only under rather special conditions, to be described in the next few paragraphs.

At first glance, one might expect that the graphite structure could simply be compressed along the c-axis like an accordion to collapse into the diamond structure. However, the requirement that the stacking arrangement change from ABAB . . . to ABCABC . . . is a serious obstacle to the formation of cubic diamond, and many experimenters have squeezed on cold graphite until they were blue in the face without recovering a single crumb of diamond. Indeed, Drickamer (1967) reported that at 25°C and 180 kbar the measured volume of graphite is about 84 per cent of its 1 atmosphere volume, and even at 300 kbar the calculated volume is still about 82 per cent of the 1 atmosphere volume; the volume of (uncompressed) diamond would be about 64 per cent of this, as noted earlier.

3.2   HEXAGONAL DIAMOND FORMATION

At second glance, the prospects for compressing graphite into hexagonal diamond might appear more favourable because both have the ABAB . . . stacking. However, a close comparison of the two structures in Fig. 2 shows that in graphite half the atoms in any plane lie directly above half the atoms in the plane below and directly below half the atoms in

the plane above, i.e. straight strings of atoms run vertically along the
c-axis, with a string atom in every plane. In hexagonal diamond, this
is not so; no strings are straight; they are all crooked. This consideration
rules out an accordion-like compression of graphite along the c-axis
to form hexagonal diamond.

However, when relatively well-crystallized graphite is compressed
at 25°C, its electrical resistivity increases markedly around 169 kbar,
especially in the a-direction, an event which suggests the formation of
new bonds between carbon atoms so as to tie up electrons formerly
available for conduction (Drickamer, 1967). However, the graphite
returns to its original state after the pressure is released. Bundy and
Kasper (1967) heated the compressed graphite to about 1000°C at
160 kbar and recovered hard, grey material which X-ray diffraction
revealed to be mostly hexagonal diamond. The crystallographic relationships between the graphite and the diamond produced from it show
that the hexagonal diamond forms by collapse of the graphite along an
a-axis, i.e. in a direction parallel with the sheets of atoms and not like
an accordion. In well-crystallized graphite, the crystallites are relatively
thick and consist of many large sheets nicely stacked so that the resulting
hexagonal diamond crystals are big enough to identify by X-ray diffraction. In poorly crystallized graphite, the nicely stacked crystallites are
too small to permit detectable amounts of hexagonal diamond to form.
The heating to 1000°C is not sufficient for melting but develops sufficient
atomic motion to build crystallites big enough to resist decompression
forces.

Hexagonal diamond has been found in meteorites (Hanneman et al.,
1967), probably as the result of impact-generated shocks on pre-
existing graphite, and also in diamond powder produced by the duPont
shock-quench process (Trueb, 1968; Balchan and Cowan, 1971). In
this process, nodules of well-crystallized graphite lie dispersed in copper
or iron; shock compression develops high pressures in the metal which
are exerted on the graphite; during decompression, the hot diamond
(and graphite) are quenched by the relatively cooler metal and the
diamond powder is eventually recovered by chemical cleaning.

3.3   CUBIC DIAMOND FORMATION

Turning now to the formation of cubic diamond from pure graphite,
the structural relationships indicated in Fig. 2 dictate a considerable

rearrangement of atoms. Experience in the manufacture of graphite indicates that the recrystallization of carbon is relatively slow for temperatures below about 2000°C. And many diamonds will survive several minutes at 2000°C in vacuum without changing into graphite. Thus, one would desire temperatures in this range and higher for changing graphite to diamond in reasonable time intervals, and, according to Fig. 1, pressures of 62 kbar or more would be needed. If the transformation to diamond involved an intermediate state which was more voluminous than the initial state, then pressure would retard the process and even higher temperatures would be necessary, but it turns out that the initial state need not be perfectly crystalline graphite, nor is the intermediate state necessarily so voluminous.

DeCarli and Jamieson (1961) first reported the recovery of diamond from medium-density graphite which had been compressed by an explosively driven shock wave. The estimated conditions for the transformation were 300 kbar and 1500 K for a few microseconds, followed by decompression. The recovered material contained a few per cent of cubic diamond in the form of agglomerates of very tiny (100 Å) crystallites. Highly crystalline graphite produced no diamond by this process, presumably because its greater density precluded the generation of sufficiently high local temperatures in the shock-compressed state.

Later, Bundy (1963) formed diamond from graphite statically compressed to pressures of about 130 kbar and heated to about 3300 K for a few milliseconds by a pulse of electric current. The diamond was recovered as a coherent mass of tiny crystallites, 200–500 Å in size. Figure 3 shows a graphite bar whose central portion has changed into diamond by this process.

According to Fig. 1, the transformation occurs at conditions which correspond to the metastable melting of graphite; the slope of this melting curve indicates that the liquid has a higher density than the solid. Thus, the route from graphite to diamond would not involve states which are more voluminous than graphite, and hence pressure should not hinder the reaction but rather should favour it. Also, according to Fig. 1, diamond should form by freezing molten carbon at pressures above about 140 kbar, the triple point. However, so far the products of such liquid quenching experiments always contain both tiny diamonds and radially crystallized flakes of graphite. One may suppose that either the local pressure falls during the freezing process so that graphite becomes stable, or else the rapid freezing does not permit

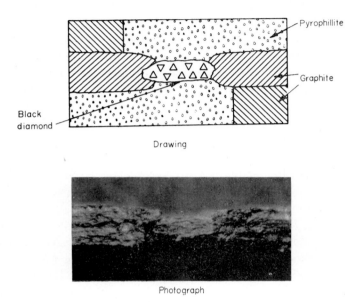

Drawing

Photograph

Fig. 3. Polycrystalline diamond formed in the central portion of a graphite bar at
130 kbar, 3300°C.

all of the carbon to crystallize as diamond. (Order in 2 dimensions
is easier to achieve than order in 3 dimensions.) No glassy carbon has
appeared despite the short quench times of a few milliseconds.

Wentorf (1965a) studied the extent of diamond formation from various
forms of non-diamond carbon during relatively long time intervals,
seconds or minutes, at pressures of 130–150 kbar and temperatures in
the range 1400–3000 K. A relatively pure highly crystalline graphite
powder changed to diamond more rapidly and completely than did
ordinary spectroscopic carbon, which is less perfectly crystalline. On
the other hand, several kinds of carbon blacks also changed more
rapidly than the spectroscopic carbon. Activation energies were estimated to be about 50–70 kcal per gram-atom, comparable with the
energy required to break two carbon–carbon bonds. Evidently certain
(unknown) structural arrangements of the atoms in the various carbon
starting materials affect the rate and completeness of the transformation
to diamond at these temperatures well below the melting point. As
remarked earlier, these are some of the high-pressure counterparts of
the diverse behaviour of various carbons and chars so well studied at
one atmosphere.

3.4 PYROLYSIS OF CARBONACEOUS MATERIALS

The effect of the structure of the carbonaceous starting material on its transformation to diamond was studied further by exposing a number of different organic compounds to 2000°C or more for 15 minutes at about 150 kbar (Wentorf, 1965a). Some of the salient results were: (1) condensed aromatic ring molecules such as naphthalene or anthracene, or nitrogen-rich compounds such as hexamethylene-tetramine or diphenylamine or phthalocyanine, produced crystalline graphite, except at higher temperatures where the initially formed graphite changed to diamond; (2) a slight departure from complete aromaticity, such as in fluorene or pentaphenylcyclopentadiene, favoured diamond formation at lower temperatures, along with some graphite; (3) aliphatic compounds such as paraffin or camphor never formed graphite but instead collapsed to diamond, often via a soft, white intermediate state whose only crystalline phase (by X-ray) was diamond. The formation of diamond from anthracene was greatly aided by the addition of say 10 per cent of camphor. Obviously seeding or nucleation effects are important in these solid-state reactions, which probably proceed by a number of complicated paths.

4 Carbon in solution

4.1 TYPES OF SOLVENTS

The transformation of graphite into diamond ought to take place more easily at lower temperatures and pressures if one could separate the carbon atoms, not so much by thermal agitation as by chemical action in some sort of a solution.

One might arbitrarily group carbon solvents into three main classes, according to the sign of the charge on the carbon atoms:

1. Negative, as in the alkaline carbides such as Li_2C_2, CaC_2, Be_2C, CeC_2, etc., some of which contain C_2^{2-} ions for example.

2. Approximately neutral, as in certain hydrous silicates and in other molten compounds, to be described soon below.

3. Positive: (a) extremely so, as in the carbonates, from which carbon can be recovered only by the action of strong reducing agents; (b) mildly positive, as in the carbides of the transition metals, from many of which carbon ex-solves by gentle means such as cooling, e.g. as from Fe_3C.

The interactions of carbon with solvents from all three classes were studied in the work which ultimately led to reproducible diamond synthesis as reported by Bovenkerk *et al.* (1959). Wentorf (1966) later made further studies on some aspects of carbon in solution at high pressures.

4.2 GRAPHITE-PRODUCING SOLVENTS

In regard to classes 1, 2, and 3(a), the current situation is that solid carbon may be liberated from them at diamond-stable pressures and temperatures such as 60 kbar and 1400–1600°C—conditions which are somewhat below those necessary for the direct transition from graphite as described in section 3. However, the liberated carbon always appears as graphite, even in the presence of diamond seed crystals.

A typical high-pressure cell for such experiments is sketched in Fig. 4. An electric current heats a spectroscopic carbon tube; the heat flow can be adjusted to create appropriate temperature gradients for carbon transfer in the carbon solvent held in the tube. The use of direct current permits one to observe the sign of the charge on the dissolved carbon. The carbon tube and its end plugs are available as carbon sources, and usually carbon could be dissolved in the hotter regions near the midlength and ex-solved in cooler regions near the ends of the tube; detectable amounts accumulated in times of 10–20 minutes.

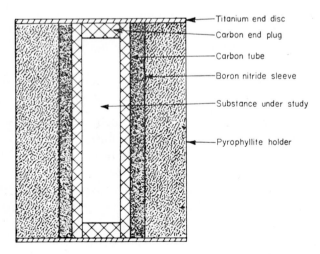

Fig. 4. High-pressure reaction cell for studying the solubility and transport of carbon.

Substances such as Li_2C_2 or CaC_2 could be made to liberate carbon at the $(+)$ pole by electrolysis; the carbon took the form of graphite. The extraction of Li from Li_2C_2 by chemical means is also favourable for producing graphite.

Many substances which are relatively volatile or unstable at ordinary pressures can be studied as carbon solvents at high pressures. Among those studied were $AgCl$, Cu_2O, $CuCl$, Cu_2S, $AlCl_3$, ZnO, ZnS, CdO, FeS, and siliceous melts containing hydroxyl such as serpentine, muscovite, biotite, and hydrous alkaline alumino-silicates. (Some natural serpentines contain dispersed graphite.) Only about 1 per cent or less of carbon appeared to dissolve in such solvents at 1500–1600°C. The ex-solved carbon always appeared as graphite. Electrolysis experiments indicated that the dissolved carbon bore negligible charge.

$CaCO_3$ did not appear to dissolve any carbon at all, but electrolysis of the melt at high pressure and temperature produced finely dispersed carbon at the cathode, presumably from local reduction by liberated calcium metal, and gas bubbles at the anode.

4.3 DIAMOND-PRODUCING SOLVENTS

These are arbitrarily named "catalyst-solvents" to distinguish them from the solvents discussed in section 4.2. They enjoy use in the commercial production of diamond from graphitic carbon at pressures in the range 40–65 kbar and temperatures of 1350–1600°C, conditions which can be economically attained in equipment of practical size. These catalyst-solvents consist of the metals such as Mn, Fe, Co, Ni, Pd, Pt (Group VIII), and alloys of these metals.

These metals are well known to be good solvents for carbon at one atmosphere and they retain this property at high pressures. For example, Fig. 5 shows the Ni-C phase diagram for 57 kbar, as found by Strong and Hanneman (1967). Here the middle portion of the composition range has been omitted in order to show the nickel-rich and carbon-rich ends more plainly. The nickel-carbon eutectic has a higher melting temperature and dissolved carbon content than at one atmosphere, and either graphite or diamond may be in equilibrium with the carbon-saturated melt, depending on the temperature. Naturally the unstable phase is slightly more soluble. At the carbon-rich end, the equilibrium temperature between pure graphite and diamond is about 1580°C. However, graphite can take up a small amount of nickel and thereby

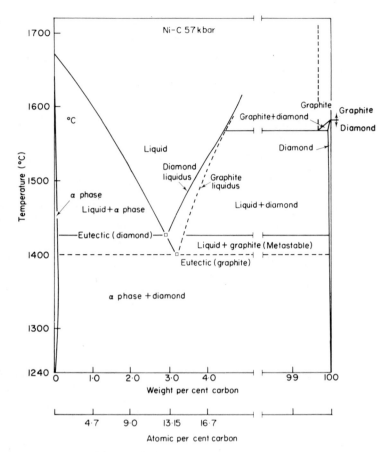

Fig. 5. Ni–C phase diagram at 57 kbar.

becomes slightly more stable than pure graphite versus diamond, and so the equilibrium temperature between diamond and graphite which has been in intimate contact with nickel is slightly lower, about 1570°C. Thus in order to be certain that diamond will crystallize from a nickel-carbon melt, the melt must be below the temperature at which the nickel-bearing graphite is stable. High pressures favour the stability of Ni_3C (Strong, 1965) and it can be recovered by quenching, but it does decompose at lower temperatures. Graphite is the product of this decomposition, even at pressures as high as 130 kbar (Wentorf, 1966).

The 55 kbar iron-carbon phase diagram resembles the nickel-carbon diagram in many ways, with the additional factor that Fe_3C is stable at all temperatures below its melting point. Thus it is possible to cool an iron melt saturated with carbon at temperatures not too far above the eutectic and obtain Fe_3C, but no diamond, in the solid. Or one may barely melt iron with diamond at high pressure and consume diamond until the iron solidifies as Fe_3C (Strong and Chrenko, 1971).

The foregoing remarks illustrate the forwardness of graphite and the shyness of diamond even where diamond is thermodynamically stable.

The admixture of nickel to iron suppresses the formation of Fe_3C, and such mixtures can form diamond at temperatures below those suitable for pure iron or pure nickel.

One might hope to find alloys of the catalyst metals—Fe, Ni, etc.— with other lower melting metals such as Cu, Si, Ge, Sn, etc. which would form diamond at relatively low temperatures and consequently lower pressures. A fair amount of work has been done in this area, for obvious economic reasons. It is found that diamond formation may proceed with dilute (e.g. 10–20 per cent Ni) alloys, but near the minimum active temperatures, say around 1100–1200°C, the diamond forms rather slowly and has a dark, porous structure of limited commercial value. Apparently some mobility of the surface atoms is required on a growing diamond crystal, and the refractory nature of carbon makes itself felt at low temperatures despite the presence of molten metal.

A simple way to transform graphitic carbon into diamond is to arrange the carbon and metal in a high-pressure reaction cell as shown in Fig. 6. After the cell has been brought up to a pressure of say 55–60 kbar, the carbon and metal are electrically heated so that the metal melts and begins to dissolve carbon. If the temperature is not too high, diamond will be more stable than graphite, and will spontaneously crystallize quite close to the graphite supply; often a film of metal less than $0 \cdot 1$ mm thick separates the two forms of carbon. If the pressure and temperature remain appropriate, the entire mass of graphite may be transformed into a mass of diamond crystals in times of the order of minutes. These growth rates correspond to a diffusion constant of about 2 to 4×10^{-5} $cm^2 s^{-1}$ for carbon in a liquid metallic phase, which seems reasonable. The diamonds may be freed from associated metal and traces of graphite by acid treatment and then sorted according to size and shape for use as industrial abrasive. Appropriate control of pressure, temperature, and catalyst metal results in crystals having particular mechanical characteristics, such as friability, for particular applications. Bovenkerk (1961) has described some of these effects.

It is not necessary to begin with pure carbon in order to make diamond. One can use carbon-bearing materials such as coal, coke, pitch, wood, peanut butter, etc. and pyrolyse them *in situ* at high pressure to provide the carbon for diamond.

If one interrupts the growth by cooling the system before all the graphite has changed to diamond, the faces of the diamond crystals

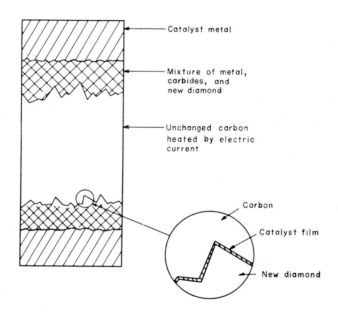

Diamond growth from graphitic carbon

Fig. 6. High-pressure reaction cell for diamond synthesis.

often have interesting features. Figure 7 shows the cube face of a diamond crystal grown from a nickel melt. The pattern of ditches and ridges arises from the gradual solidification of the melt. The cube face of diamond bears a fairly high density of free carbon valence bonds: two per atom. The lattice constant of nickel is close to that of diamond, so that a cube diamond face exerts a strong orienting effect upon nearby nickel as it freezes. The freezing begins at separate patches on the surface; under these frozen patches the growth of diamond is negligible, but carbon continues to be deposited from the molten metal between patches so as to build up rather low (1 μm) ridges of diamond before the entire surface becomes frozen. On an octahedral diamond face the free carbon bond density is much lower, the orientation effects upon nickel are much weaker, the freezing is more random, and the resulting patterns of ridges are more tortuous.

Fig. 7. Cube face of diamond grown by nickel catalyst.

If, in the experiment depicted in Fig. 6, one uses direct current for heating, differences in diamond growth appear, as shown in Fig. 8 (Wentorf, 1966). More diamond appears at the carbon–metal interface where the carbon is more positive. This implies that the carbon dissolved in the molten catalyst metal bears a net (+) charge so that it is more rapidly driven away from the graphite through the metal film to become diamond. At the opposite metal face, the movement of carbon is hindered by the electric field. This view of the charge on the dissolved carbon is consistent with the chemical nature of carbonyls and with low-pressure electrolysis experiments on solid solutions of carbon in metals (Kovenskii, 1965). One may also note that positive ions are usually smaller than negative ions of the same element, and that diamond is more densely packed than graphite. The data of Strong (1964) on the solubility of carbon in molten nickel at high pressure indicate that there the carbon has a partial atomic volume of about

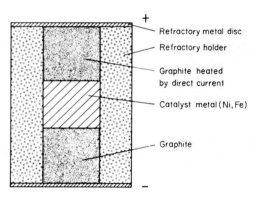

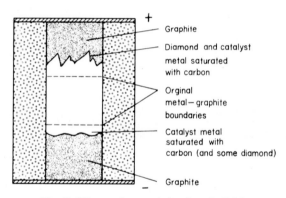

Fig. 8. Diamond growth in electric field.

5 cm³ per gram-atom, compared with graphite at 5.4 cm³ per gram-atom and diamond at 3.4 cm³ per gram-atom. Thus the $p\Delta V$ contribution to the free energy would act to discourage graphite formation from molten nickel, but where the dissolved carbon has more negative charge and is consequently larger, graphite formation would not be so discouraged.

From arguments about the chemical nature of dissolved carbon such as these, one might hope to make a diamond-forming catalyst from an alloy of a strong carbide-forming metal, such as Nb, V, or Ti, with atoms such as Cu which have enough electrons to complete the octet involved in carbonyl formation. In other words, perhaps the "atomic average" of Ti and Cu would resemble Fe in behaviour with carbon. Experiments by Wakatsuki (1966) show that small amounts of diamond may indeed be formed with such alloys as molten catalysts, although the required temperature and pressure are significantly higher than required for Fe. Evidently the electronic states of individual metal atoms are at least as important for diamond formation as the average electronic abundance.

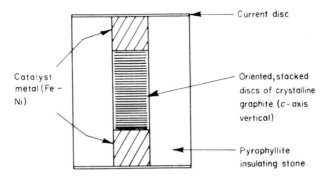

Fig. 9. High-pressure reaction cell using oriented crystalline graphite for carbon source.

It is not always necessary for the carbon to be thoroughly dispersed into separate atoms in order for diamond to form. Wentorf (1965b, 1966) has described experiments using oriented graphite pieces for the carbon supply, as shown in Fig. 9. The graphite pieces were cut from highly crystalline natural graphite flakes and were aligned during stacking up in the cell by means of the characteristic striae on the flakes. These striae are oriented relative to the carbon hexagons as shown in Fig. 10. Figure 11 shows a few octahedral faces of diamond grown relatively rapidly from this graphite at pressures several kbar above equilibrium. Some unchanged graphite flakes, with striae, still adhere to the diamond faces, and the orientation of the striae relative to the diamond faces forces one to conclude that the diamond formed out of the graphite by a simple accordion-like collapse along the c-axis, with the orientation relationship shown in Fig. 10 still preserved. This could happen only if the diamond were built up out of such large sheets

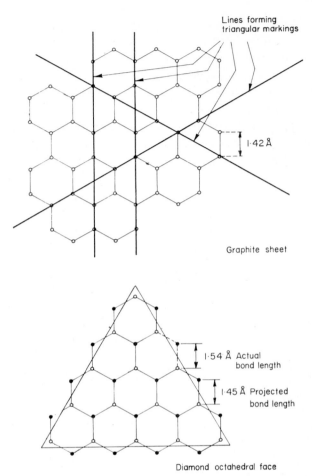

Fig. 10. Carbon atom networks in diamond and crystalline graphite and relationship to graphite striae.

of carbon hexagons that the sheets were unable to rotate on the way across the metal film from graphite to diamond. However, the molten catalyst metal medium permits enough freedom for the building up of the ABCABC... layer sequence of diamond from the ABAB... sequence of graphite.

If the pressure is reduced so that the rate of diamond growth is not so rapid, then the oriented relationship between source graphite and diamond fades away, and probably smaller carbon particles are involved in the transformation to diamond under these conditions.

Fig. 11. Diamond octahedral face covered with thin layer of crystalline source graphite, showing striae.

On the basis of such experiments, one might hope for diamond-forming action from the compounds or elements which intercalate graphite, such as H_2SO_4, $CuCl_2$, Na, etc. So far experiments along these lines have not made any diamond. Sodium will form some carbide at moderate pressures; $CuCl_2$ does not appear to be as good a carbon solvent as CuCl. The swelling ordinarily associated with the intercalation of molecules would hinder this reaction at high pressures (1 cm^3 at 40 kbar is about 2 kcal of free energy). On the other hand, the "intercalation" of nickel atoms to dissolve the carbon results in a slight decrease in volume, as noted earlier.

5 Growth of large laboratory diamond crystals

In the growth of diamond from graphite as described earlier, the driving force is the free energy difference between diamond and graphite which produces a solubility difference; the larger this difference, the greater the rate of sponataneous nucleation and the smaller the crystals. In order to grow large crystals, it is necessary to maintain a small free energy difference for a long time, but several difficulties arise on the way from graphite to diamond. The most serious of these is the precise control of temperature and pressure at the growing diamond interface, where these variables are not only extremely difficult to measure to the required accuracy, but also suffer erratic changes due to recrystallization, etc. during the growth period.

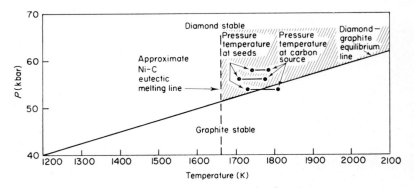

Fig. 12. Portion of carbon phase diagram showing temperature intervals for carbon source and growing diamond seeds.

However, if the free energy difference is supplied by the carbon solubility difference arising from a temperature difference, the crystallization ought to be easier to control because it is experimentally easier to maintain a small temperature difference than to maintain precise values of temperature and pressure simultaneously. Figure 12, from Wentorf (1971), depicts this idea. Carbon is supposed to dissolve from a source of carbon held at some high temperature and travel through a molten catalyst metal bath (Ni, Fe, etc.) to crystallize as diamond at a slightly lower temperature. In principle, it matters little whether the carbon supply is graphite or diamond. In practice it is easier to operate at pressures at which all carbon is stable as diamond. Then the bothersome volume change associated with transformation to diamond does not occur during the main growth period.

Figure 13 shows a high-pressure reaction cell arranged for diamond growth by this method, as described by Wentorf (1971). The heat generation and flow in the cell here make the midlength regions warmer than the ends. Thus the carbon supply might be held at about 1450°C while the diamond seed crystals would be held at about 1420°C, all at a pressure of about 55–60 kbar. The carbon supply begins as a mixture of diamond grit and graphite so as to obtain as high an initial density as possible, but the graphite changes to diamond in the first few minutes after the catalyst bath metal melts. The seed crystals are small diamonds; other substances such as Al_2O_3, SiC, BN, MgO, or graphite are not effective as diamond seeds. If no diamond seeds are present, graphite usually crystallizes instead of diamond.

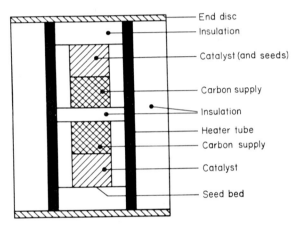

End disc
Insulation

Catalyst (and seeds)

Carbon supply
Insulation
Heater tube
Carbon supply

Catalyst

Seed bed

Fig. 13. High-pressure reaction cell for growth of diamond crystals in temperature gradients.

The usual bath metals can hold more carbon in solution at high pressures than at one atmosphere, so as a rule the diamond-growing bath is not saturated with carbon when it first melts. The required carbon comes from the source and also partly from the seeds, so that the seeds suffer some etching and dissolution. The seeds are held mechanically in the cool ends. In the top end, convection helps to stir the bath, and also tiny diamond crystals may float up from the carbon supply to act as seed crystals, so that usually many smaller crystals grow in the cool top end. If one desires only one or just a few crystals, they are grown in the cool end of the bottom bath where things are a little calmer. The carbon flux in the bottom bath is about 10^{-4} g s^{-1} cm^{-2}. If the carbon flux is too high, the growing crystals cannot accommodate it as high-quality growth, and in addition graphite may nucleate and grow.

Figure 14 shows a diamond crystal, grown from a single seed, partially embedded in the frozen bath metal; the metal has been partly cracked away. The fresh crystal faces are fairly smooth so that it is easy to see the interior of the crystal, but a microscopic examination of the faces often reveals low steps or layers of growth, as shown in Fig. 15. When impurities, such as bath metal, are trapped in the crystals, they

Fig. 14. Single crystal of diamond half-buried in the frozen catalyst metal bath.

usually take the form of thin platelets parallel with the prominent faces. Thus the crystals appear to grow layer by layer, with new layers beginning on edges or corners (which represent vestiges of the fastest-growing faces) and spreading across the slowest-growing faces such as ⟨111⟩ and ⟨100⟩. The layers appear to be a few microns thick.

The geometry of advancing layers of new growth indicates that the fresh, exterior corner will receive the greatest flux of carbon atoms, while at the interior corner, next to the surface being covered, the concentration of carbon is lower and impurities recently detached from the surface are abundant, as sketched in Fig. 16. The higher the carbon

Fig. 15. Crystal face of freshly grown diamond showing layers of growth.

flux to the crystal, the thicker is the advancing layer, and the more pronounced are these concentration differences, which in themselves favour thicker layers or roofing over of pockets of impurity in the crystal. A thin layer under a high carbon flux will catch up to a thicker layer, which moves slower on account of the confusion or lack of carbon at its inside advancing corner, so as to make a new, thicker layer which moves even slower if it is to form good crystal. Thus the potential for

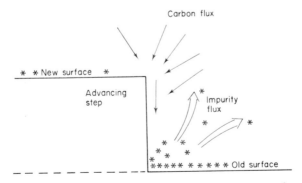

Fig. 16. Fluxes of carbon and other matter near a growth step.

growing sound crystals deteriorates rapidly once a critical carbon flux is exceeded; new layers are nucleated too fast and rapidly coalesce to form thicker layers which are prone to trap impurities under them.

To ensure sound growth, one must wait for each layer step to move far enough away from the layer nucleation area so that the next layer to appear will not coalesce with it on its sweep across the face, despite retardation at inevitable difficult regions on the growing face. In effect, this means that the layer nucleation rate should be inversely proportional to the face length, the distance which a layer has to travel.

If all layers are about the same thickness and move at about the same speed when the crystal is growing at the maximum possible rate for depositing sound material, then the average rate of crystal size increase, r mm h^{-1}, should be related to the instantaneous average crystal size, S by

$$2srb = 1$$

where b is a parameter related to the particular conditions of growth. In this work the experience with many crystals put the value of b around 2.5 h mm^{-2}, although it varies somewhat for different baths.

Then the time θ required to grow to a size L at the maximum rate for sound crystals would be

$$\theta(L) = \int_0^L \mathrm{d}s/r = \int_0^L 2bs\,\mathrm{d}s = bL^2.$$

Thus one estimates times of 90 h for 6 mm crystals. In practice, it is difficult to adjust the growth rate continuously according to the instantaneous crystal size. Instead, one is forced to grow most of the crystal at the rate appropriate for the final size. Thus a 5 mm (1 carat, 0.2 gram) high-quality crystal requires about a week to grow. Such growth rates are not economically competitive with digging natural diamonds out of the ground, but the close control over the growth process permits the study of some of the properties of diamond crystals and the effects of impurities on these properties. Strong and Chrenko (1971) have done considerable work in this area.

6 Impurities in diamonds

A growing diamond crystal may incorporate other substances in various states of aggregation. Natural diamonds are known to contain varying amounts of nitrogen either as platelets or as dispersed atoms; they

frequently contain inclusions of various minerals such as olivine, diop-
side, garnet, etc. which have been studied in considerable detail (Meyer
and Boyd, 1968; Milledge, 1970).

As noted in the discussion of the growth of large crystals, special
pains are necessary to prevent the trapping of pockets of catalyst metal
melt inside synthesized diamond crystals, even those which are not so
large. Most synthesized diamonds used for industrial grit contain some
inclusions, usually metallic. For example, one may detect nickel in
diamond crystals grown in a nickel system. The trapped nickel has a
Curie temperature similar to that of ordinary nickel. Some stable
carbides, such as TiC or Cr_3C_2, may co-crystallize with diamond and
can be seen as needles or plates inside synthesized diamond crystals.
The Cr_3C_2 needles often lie oriented so that the strings of C atoms in
the carbide are parallel with the (110) strings of C atoms in the diamond.

The diamond lattice is rather small and will accommodate only a few
kinds of foreign atoms, such as Li, Be, B, N and Al, when they are
present in the catalyst metal melt. The four metals mentioned make the
resulting diamond a p-type semiconductor (Wentorf and Bovenkerk,
1962), with boron imparting a blue colour. The growth of "large"
(i.e. 5 mm) laboratory diamond crystals has permitted more investigation of the effects of foreign atoms (Strong and Chrenko, 1971). Nitrogen
and boron have been studied the most.

Nitrogen usually enters synthesized diamond as single atoms which
randomly substitute for carbon. They give rise to a characteristic optical
absorption in the blue region so as to give the crystal a yellow colour at
concentrations of 10^{17} atoms per cm^3. At higher nitrogen concentrations,
about 3×10^{19} atoms per cm^3, or one nitrogen for every 2000 carbon
atoms, the colour deepens to a deep golden yellow, and still higher
concentrations result in greenish hues. It is difficult to achieve higher
nitrogen concentrations without blocking diamond growth completely.
Evidently growing diamond takes up nitrogen rather readily and the
usual carbon sources and catalyst metals contain enough available nitrogen to impart a yellow colour. Specially purified growth media are
needed to grow colourless crystals.

The nitrogen atoms cause the diamond to absorb in the ultraviolet
at wavelengths longer than the 2250 Å absorption characteristic of the
pure lattice bandgap at $5 \cdot 4$ eV and also in the infrared at $8 \cdot 85$ μm.
The uncompensated electrons result in a paramagnetic spin resonance
but not in semiconducting behaviour. The energy level of the extra

electron of the nitrogen lies about 4 eV below the conduction band, too far removed for appreciable conduction at ordinary temperatures. In fact, n-type semiconducting diamonds are unknown in nature and can be made so far only by ion implantation, not by doping during growth.

Natural diamonds may contain up to 0.1 per cent of nitrogen (Kaiser and Bond, 1959), sometimes as isolated nitrogen atoms; but more frequently the nitrogen is gathered together as platelets 600 to 100 Å across, lying in the ⟨100⟩ planes. Such diamonds are usually colourless but absorb in the ultraviolet like the yellow crystals and in the infrared at 7.8 μm. It is not clear whether the platelets form by an annealing-recrystallization process, whether the original carbon carried nitrogen in this form, or whether some other mechanism was responsible. They are never found in synthesized diamonds. The platelets affect the physical properties of diamond (Berman, 1965) by interfering with the transport of heat, and by impeding the motion of dislocations at high temperature and mechanical stress (Evans and Wild, 1965).

By contrast, dispersed nitrogen appears to weaken the crystal slightly. However, it barely affects the thermal conductivity, which remains about four times that of copper (Strong and Chrenko, 1971). The nitrogen is not always taken up uniformly in the growing crystal but instead may appear as streaks or bands of yellow mixed with colourless regions when the overall nitrogen content is not too high.

Boron enters diamond rather easily, either by diffusion into existing crystals or by uptake of boron from a catalyst metal bath during growth (Wentorf and Bovenkerk, 1962).

If diamond crystals are held for 10 minutes at say 60 kbar and 1600°C in a boron-rich environment, such as a mixture of B_4C and graphite or inert metals containing boron, the surface of the diamond acquires tints of blue and exhibits p-type semiconductivity. The conductivity may become almost metallic in character with increasing exposure to boron. Electron diffraction patterns of exposed surfaces may show an extra set of spots which can be interpreted as arising from boron atoms residing in the natural interstices of the diamond lattice—these interstices are large enough to accept boron atoms with little strain.

Growing diamond will also incorporate boron atoms. Diamond made directly from boron-doped graphite by flash-heating to 3300°C at 130 kbar is a p-type semiconductor (Bundy, 1963, 1969). When the diamond is grown using a molten catalyst metal, boron is taken up from the bath. A few parts per million of boron in the diamond produce perceptibly blue, p-type semiconducting diamonds. The boron may be taken up more strongly in certain crystal directions than others to produce the coloured sectors shown in Fig. 17. At large boron concentrations, e.g. 100 ppm, the crystals become an opaque blue-black with high, almost degenerate conductivity. The boron atoms replace carbon and in addition to the carbon bands found in pure diamond at 2 to 6 μm, there appears a characteristic optical absorption at 3.5 μm,

Fig. 17. Section of boron-doped laboratory-grown diamond crystal showing sectors of colour due to uneven incorporation of boron.

corresponding to an activation energy for conduction of 0.373 eV, in fair agreement with electrical measurements. Messmer and Watkins (1970) have made successful theoretical treatments of the effects of boron or nitrogen substituents in diamond. Chrenko (1971) has described methods for analysing boron in diamonds by activation techniques.

Some boron-doped diamond crystals have been used as thermistors which are stable at temperatures up to 600°C.

Atoms of Al, Be, or Li, when added to diamond-growing systems, also produce semiconducting diamonds with activation energies for conduction in the range 0.2–0.35 eV, but these systems have not been studied as extensively as those with boron or nitrogen. Oxygen and hydrogen may be detected in the products of vacuum fusion of diamond with pure Fe, but very little is known about the role of these elements in diamond. The extreme mobility of hydrogen at high temperatures and pressures makes the study of its effects rather difficult. Oxygen may be added to diamond-growing systems in the form of Fe_3O_4, NiO, etc. and some CO is thereby formed. Oxygen which is chemically saturated by being tightly bonded to diamond surfaces is largely responsible for the low friction coefficient of diamond and its reluctance to be wetted except by extremely active metals such as Ti.

7 Polycrystalline diamond masses

Diamond crystals cleave rather easily along the $\langle 111 \rangle$ planes, where the bond density is lowest. This kind of friability restricts the use of diamond tools in many applications, particularly where they may encounter abrupt mechanical stresses. Certain types of natural diamond found in Brazil, called carbonado and ballas, are polycrystalline aggregates of many small diamond crystals, randomly oriented. Such aggregates are less easily broken than single crystals and enjoy usage in special applications. Close examination of such masses reveals that the crystals are bonded to each other primarily by carbon-to-carbon bonds (Trueb and Butterman, 1969). The masses retain their integrity after prolonged treatment with acids which dissolve away foreign material between the grains. Carbonado may contain up to 20 per cent by volume of impurity, usually oxides of iron and silicates, but ballas contains very little non-diamond material. Man's limited knowledge of natural diamond formation has so far not permitted a good description of the processes by which carbonado or ballas may form, whether by bonding together of existing small grains, or by crystallization of an originally amorphous mass of some carbonaceous material, or by a direct transformation of graphite into a diamond mass, as done by Bundy (1963). Evidently the conditions required are rather special because such masses, with extensive diamond-to-diamond bonding, are found only in Brazil and to a lesser extent in West Africa.

Of course it is possible to make a mixture of diamond grit and metal in which the diamonds are held in place mechanically or by carbon-to-metal bonds. Such bodies are limited to about 70 per cent diamond by volume and may be prepared at high or low pressures using a variety of metals or alloys, among the most popular of which are Cu, Ti, Ta,

W, WC, Co, etc. Such bodies generally fall short of the goal of a carbonado-like material because the lack of true diamond-to-diamond bonding results in inferior thermal conductivity and strength, particularly at higher temperatures.

Several experimental difficulties become apparent when one attempts to synthesize masses like carbonado in the high-pressure laboratory. Firstly, if diamond sintering is desired, one might plan on temperatures of about 2/3 of the melting temperature, so that temperatures upward of 2200°C are involved. A look at the carbon phase diagram, Fig. 1, indicates that pressures of at least 70 kbar will be required if the diamond is not to turn completely into graphite. Second, the strength of diamond, even at high temperatures, makes it difficult to squeeze all the empty spaces out of a mass of diamond crystals. If the average pressure on such a mass is say 70 kbar, the pressure on bearing points of diamond crystals will be much higher but the pressure on free surfaces not in contact will be quite low, and if the temperature is high enough the diamond will change to graphite on these surfaces. Third, the persistence of the intercrystalline voids may permit undesirable foreign material to squeeze in amongst the crystals. Fourth, no matter what the diamond mass is embedded in, the superior elastic stiffness of diamond ensures that upon pressure reduction the greater expansion of the embedding material will exert tractive forces on any diamond mass, to try to pull it apart into smaller pieces. In spite of these difficulties, some progress has been made.

Stromberg and Stevens (1970) operated at temperatures of 1800–1900°C and pressures of 60–65 kbar for times of 1 hour. They cleaned fine (0.1–10 μm) diamond powder very carefully with various liquids, and then heated it in high vaccum before sealing it into tantalum crucibles by electron-beam welding in high vacuum. The sealed crucibles could be enclosed in indium to reduce breaking up on pressure reduction. Selected impurities such as B, Be, or Si were sometimes added to the diamond powder to improve sintering, and they had a beneficial effect. Some of the compacts produced had densities and hardnesses not too far from those of single crystal diamond. Others contained soft spots, the result of inadequate sintering. X-ray diffraction studies of the compacts indicated that some Ta and Si (when present) had formed carbides.

Bovenkerk et al. (1964) employed boron carbide as a binder and promoter of carbon migration to form polycrystalline masses of diamond at pressures of 30 to 65 kbar and temperatures in the range 1200–1700°C. The higher pressures and temperatures produced stronger bodies.

Hall (1970) has described procedures for bonding small diamond particles together to form carbonado-like masses. The diamond powder was held in a graphite tube and exposed to pressures in the range 65–85 kbar and temperatures of 2400–2500 K for periods of a few minutes. Hall indicates that at the higher pressures a strong, white product having a density of about 99 per cent that of pure diamond was produced. At 65 kbar, diamond is theoretically not stable at the temperatures mentioned, but if the heating period is kept to about 20 seconds, only a portion of the diamond changes to graphite and the product is relatively strong. Binding agents such as refractory carbides, borides, nitrides and oxides may also be used with the diamond powder. Sintered pieces as large as 8 mm have been made. Various shapes, including holes, can be formed to eliminate much tedious grinding.

Katzman and Libby (1971) reported on the making of sintered diamond compacts with a cobalt binder at about 60 kbar and 1600°C. Fine diamond powder (80–88 volume per cent) was mixed with cobalt powder (20–12 volume per cent) and contained in tantalum for periods of 20 minutes at the above conditions. The product masses were about 6 mm in diameter and 2.5 to 9 mm long, with nearly theoretical densities and hardnesses approximately half that of solid diamond. They performed fairly well in shaping grinding wheels. Although some diamond-to-diamond bonding appeared to have occurred, as noted by microscopic observation, masses made with less than 20 volume per cent cobalt were weaker. When some graphite powder was mixed with the diamond powder, the graphite changed into diamond in the mass, but the diamond grain size increased with a concomitant drop in overall hardness.

This process resembles that patented by DeLai (1964) in which a number of metals, such as Fe, Ni, Co, Ti, Pd, Rh, etc. and their alloys, are used to bond diamond particles together at high pressures and temperatures, e.g. 65 kbar and 1700°C, in times of about 15 minutes.

Vereshchagin et al. (1968, 1969, 1971a) have reported on the synthesis of polycrystalline diamond bodies which bear some resemblance to natural ballas and carbonado. Most of these were made from graphite by methods not disclosed in detail. Another kind, apparently made from a mixture of diamond powder and binder, was proposed as a construction material for high-pressure apparatus; lumps up to 12 mm in size were made (Vereshchagin *et al.*, 1971b). Photomicrographs of polished surfaces of all these Russian compacts show negligible diamond-to-diamond bonding; instead, the masses are held together by binders.

In spite of the progress described above, it appears that considerable improvement in the coherence of diamond compacts is required before synthetic products approaching the best natural carbonado or ballas are available. Perhaps these synthetic diamond conglomerates will be useful in extending the range of high-pressure apparatus if they can be used for the most highly stressed portions.

8   Unexplored territory

Although men now understand much more about diamond formation than they did 20 years ago, several areas of ignorance remain.

a. Most natural diamonds do not resemble synthesized diamond in many respects, for example: (1) colour (no red tints formed in the laboratory); (2) layering of growth noted in cross-sectioned and etched natural diamonds; (3) crystal habit ($\langle 110 \rangle$ faces rare in synthesized diamond); (4) nitrogen platelets, which are not found in synthesized diamond; (5) mineral inclusions.
b. Carbonado and ballas formation are not understood.
c. The melting line of diamond versus pressure is incompletely known. Perhaps a metallic form of carbon may exist, in analogy with Si, Ge and Sn. The melting temperature of diamond cannot decrease indefinitely with increasing pressure.

Perhaps the work of the next 20 years will shed some light on these problems and, more likely, will uncover new phenomena which will increase our understanding of this world and benefit mankind.

References

Alder, B. J. and Christian, R. H. (1961). *Phys. Rev. Lett.* **7**, 367.

Angus, J. C., Will, H. A. and Stanko, W. S. (1968). *J. Appl. Phys.* **39**, 2915–2922.

Balchan, A. S. and Cowan, G. R. (1971). US Patent No. 3,608,014. Issued September 21, 1971, and assigned to E.I. duPont de Nemours and Co., Wilmington, Del.

Berman, R. (1965). "Physical Properties of Diamond" (Ed. R. Berman). Clarendon Press, Oxford.

Berman, R. and Simon, F. (1955). *Z. Elektrochem.* **59**, 333.

Bovenkerk, H. P. (1961). "Progress in Very High Pressure Research", pp. 58–69. John Wiley, New York.

Bovenkerk, H. P., Savage, R. H. and Wentorf, R. H. (1964). US Patent No. 3,136,615. Issued June 9, 1964, and assigned to the General Electric Company.

Bovenkerk, H. P., Bundy, F. P., Hall, H. T., Strong, H. M. and Wentorf, R. H. (1959). *Nature*, **184**, 1094.

Bundy, F. P. (1963). *J. Chem. Phys.* **38**, 618, 631.

Bundy, F. P. (1969). *Kon. Ned. Akad. Wetensch. Proc. Ser. B*, **72**, 302.

Bundy, F. P. and Kasper, J. S. (1967). *J. Chem. Phys.* **46**, 3437.

Bundy, F. P., Bovenkerk, H. P., Strong, H. M. and Wentorf, R. H. (1961). *J. Chem. Phys.* **35**, 383.

Bundy, F. P., Strong, H. M. and Wentorf, R. H. (1972). Methods and mechanisms of growth of synthetic diamond. *In* "Chemistry and Physics of Carbon" (Ed. P. L. Walker), Vol. 7. (In press).

Chrenko, R. M. (1971). *Nature, Phys. Sci.* **229**, 165.

DeCarli, P. S. and Jamieson, J. C. (1961). *Science*, **133**, 1821.

DeLai, A. J. (1964). US Patent No. 3,141,746. Issued July 21, 1964 and assigned to the General Electric Company.

Deryagin, B. V., Lyuttsau, V. G., Fedoseev, D. V. and Ryabor, V. A. (1970). *Dokl. Akad. Nauk. SSSR.*, **190**, 86. (See also Deryagin, B. V. (October 30, 1969). *New Sci.* 228.)

Drickamer, H. G. (1967). *Science*, **156**, 1183.

Evans, T. and Wild, R. K. (1965). *Phil. Mag.* **12**, 479.

Eversole, W. G. (1961). Canadian Patent No. 628,567 (October 3, 1961).

Hall, H. T. (1970). *Science*, **169**, 868.

Hanneman, R. E., Strong, H. M. and Bundy, F. P. (1967). *Science*, **155**, 995.

Kaiser, W. and Bond, W. L. (1959). *Phys. Rev.* **115**, 857.

Katzman, H. and Libby, W. F. (1971). *Science*, **172**, 1132.

Kovenskii, I. I. (1965). *Sov. Phy.–Solid State*, **5**, 1036.

Messmer, R. P. and Watkins, G. D. (1970). *Phys. Rev. Lett.* **25**, 656.

Meyer, H. O. A. and Boyd, F. R. (1968). Geophysical Laboratory, Carnegie Institution of Washington, Year Book 1968, pp. 315–322: "Inclusions in Diamonds".

Milledge, H. J. (1970). "Inclusions in Natural Diamonds". Private communication.

Rossini, F. D. and Jessup, R. S. (1938). *J. Res. Nat. Bur. Stand.* **21**, 491.

Stromberg, H. D. and Stevens, D. R. (1970). *Ceram. Bull.* **49**, 1030.

Strong, H. M. (1964). *Acta Met.* **12**, 1411.

Strong, H. M. (1965). *Trans. Met. Soc. AIME*, **233**, 643.

Strong, H. M. and Chrenko, R. M. (1971). *J. Phys. Chem.* **75**, 1838.

Strong, H. M. and Hanneman, R. E. (1967). *J. Chem. Phys.* **46**, 3668.

Trueb, L. F. (1968). *J. Appl. Phys.* **39**, 4707. (See also US Patent No. 3,401,019, G. R. Cowan *et al.*, 1965, assigned to E.I. duPont de Nemours and Co.)

Trueb, L. F. and Butterman, W. C. (1969). *Amer. Mineral.* **54**, 412.

Vereshchagin, L. F., Nikol'skaya, I. V., Orlov, Yu. L., Feklicher, E. M. and Kalashnikov, Ya. A. (1968). *Dokl. Akad. Nauk SSSR*, **182**, 77–79.

Vereshchagin, L. F., Yakovlev, E. N., Varfolmeeva, T. D., Slesarev, V. N. and Sterenberg, L. E. (1969). *Dokl Akad. Nauk SSSR*, **185**, 555–556; (1971a). *High Temp.–High Pressures*, **3**, 239.

Vereshchagin, L. F., Semerchan, A. A., Modenov, V. P., Bocharova, T. T. and Dmitriev, M. E. (1971b). *Sov. Phys.-Dokl.* **15**, 1065.

Wakatsuki, Masa (1966). *Jap. J. Appl. Phys.* **5**, 337.

Wentorf, R. H. (1965a). *J. Phys. Chem.* **69**, 3063.

Wentorf, R. H. (1965b). "Advances in Chemical Physics" (Ed. I. Prigogine), Vol. IX, pp. 365–404. Interscience, London, New York.

Wentorf, R. H. (1966). *Ber. Bunsenges. Phys. Chem.* **70**, 975.

Wentorf, R. H. (1971). *J. Phys. Chem.* **75**, 1833.

Wentorf, R. H. and Bovenkerk, H. P. (1962). *J. Chem. Phys.* **36**, 1987.

Author Index

Numbers in italics indicate the pages on which the references are listed.

A

Adadurov, G. A., 80, 118, *146, 151*
Adams, L. H., 169, 192, 198, 233, *243, 244*
Adler, D., 99, *146*
Agron, P. A., 187, *245*
Ahrens, T. J., 18, 86, 100, 105, 110, 115, 118, 122, 125, *146, 150, 152*, 218, 231, *243*
Ainsworth, D. L., 80, *146*
Akimoto, S., 219, 221, 223, 230, 232, 237, *245, 246*
Alder, B. J., 58, 100, 118, *146*, 192, *243*, 251, 252, *279*
Aliev, Z. G., 118, *146*
Allison, F. E., 104, 105, *146*
Altman, F. D., 132, *159*
Al'tshuler, L. V., 6, 80, 97, 100, 117, 118, 119, 120, 121, 130, 145, *146, 147, 151*, 181, 203, 205, *243*
Alverson, R. C., 126, *141*
Anderson, D. H., 122, 123, *153*
Anderson, D. L., 118, *146*, 231, *243*
Anderson, G. D., 18, 39, 125, 126, 129, *141, 159*
Anderson, G. W., 52, 53, *159*
Anderson, O. L., 183, 206, 226, 227, 228, 230, 234, *243, 246*
Andreatch, P., 183, *243*
Andrews, D. J., 18, 117, *147, 154*
Angus, J. C., 250, *279*
Asay, J. R., 126, 127, 129, *147*
Atovmyan, L. O., 118, *146*
Averbach, B. L., 140, *162*

B

Backman, M. E., 25, *147*
Bailey, L., 6, *147*
Bakanova, A. A., 6, 120, 121, *146, 147*, 203, 205, *243*
Balchan, A. S., 58, 86, 98, *147*, 202, 203, 211, 213, 239, 242, *243*, 254, *279*
Bancroft, D., 51, 98, 115, *147, 152*, 207, 212, 239, *243*
Band, W., 39, 44, 126, *147, 155*
Banus, M. D., 98, *154*
Barbee, T. W., 21, 141, 142, 143, *147, 161*
Barker, L. M., 22, 46, 79, 82, 83, 84, 85, 86, 125, 126, 131, *147, 148, 155, 158*
Barnett, J. D., 169, 180, 181, 184, 220, 224, 225, 231, 237, 240, *243, 244, 245*
Barrington, J., 140, *148*
Barsch, G. R., 185, 186, 187, 188, 189, 190, 191, 192, 241, *243, 244*
Barsis, E., 67, *148*
Bartel, L. C., 35, 101, 116, *149, 158*
Bartels, R. A., 181, *243*
Bassett, W. A., 145, *148*, 169, 171, 173, 176, 179, 180, 181, 182, 183, 184, 192, 193, 194, 195, 196, 197, 198, 199, 200, 201, 203, 204, 206, 207, 208, 209, 212, 213, 214, 215, 216, 217, 219, 220, 221, 223, 224, 225, 230, 231, 232, 235, 236, 237, 240, *243, 244, 245, 246, 247*

Batsanov, S. S., 138, *148*
Baum, N., 77, *148*
Bavina, T. V., 118, *146*
Beer, A. C., 112, *162*
Belechar, T., 51, *152*
Belyakov, L. V., 122, *148*
Benedick, W. B., 52, 53, 54, 55, 56, 57, 89, 102, 103, 112, 124, 125, 126, *149, 153, 155, 156, 159*
Bennion, R. B., 169, *244*
Berard, C., 111, *150*
Berg, U. I., 58, *148*
Berger, T. L., 133, 135, 136, 138, *160*
Bergmann, O. R., 140, *148*
Berman, R., 252, 274, *279*
Bernstein, D., 58, 59, 66, 67, *148, 156*
Bezukov, G. I., 105, 106, *154, 155*
Birch, F., 187, 188, *244*
Blackburn, L. D., 208, *244*
Bland, D. R., 39, *148*
Boade, R. R., 18, 56, *148, 158*
Bocharova, T. T., 278, 279, *281*
Bond, W. L., 168, *246*, 274, *280*
Born, M., 227, *244*
Borod'ko, Y. G., 118, *146*
Bovenkerk, H. P., 252, 258, 261, 273, 274, 277, *279, 280, 281*
Bowden, H. G., 133, *148*
Boyd, F. R., 273, *280*
Boyd, T. J., 86, *162*
Bradley, R. S., 158, *244*
Brazhnik, M. I., 6, 18, 121, 130, *146, 147, 148, 157*
Bredig, M. A., 187, *245*
Breed, B. R., 86, 89, 141, *148*
Breusov, O. N., 97, 118, *146, 151*
Bridgman, P. W., 58, 97, 121, 145, *148*, 166, 168, 180, 185, 188, 190, 191, 192, 193, 194, 195, 197, 203, 205, 207, *244*
Brillhart, D. C., 134, 140, *148*
Brooks, H., 99, *146*

Brooks, W. P., 52, 53, 130, *153, 159*
Brown, N., 137, 138, *149, 157*
Brugger, R. M., 169, *244*
Bundy, F. P., 58, 116, 118, *148*, 207, *244*, 250, 252, 254, 255, 258, 274, 276, *280*
Bunker, R., 77, *148*
Burgess, T. J., 85, *155*
Burley, G., 197, *244*
Burton, B. L., 51, *152*
Butcher, B. M., 18, 19, 47, 125, 126, 127, 141, *148, 149, 159, 161*
Butterman, W. C., 276, *280*
Buzhinskii, O. I., 90, 111, *149*

C

Cable, A. J., 5, 6, *149*
Campbell, J. K., 193, 195, 197, *244*
Canfield, J. M., 54, *149*
Carpenter, S. H., 134, *149*
Carroll, M. M., 21, *149*
Carter, W. J., 8, 14, 22, 125, 129, 145, *152, 158*, 181, 182, 183, 184, 192, 196, *245*
Champion, A. R., 89, 101, 108, 121, *149*
Chang, H. L., 127, *149*
Chang, Z. P., 185, 186, 187, 188, 189, 190, 191, 192, 241, *243, 244*
Chao, E. C. T., 136, *149*
Charest, J. A., 141, *149*
Chatterjee, A., 18, *159*
Chen, P. J., 24, 25, 94, *149*
Chou, P. C., 22, *149, 161*
Chrenko, R. M., 261, 272, 273, 274, 275, *280*
Christian, R. H., 34, 58, 89, 100, 118, 144, *146, 162*, 181, 185, 186, 190, 191, 192, *244*, 251, 252, *279*
Christensen, A. B., 130, *151*
Christman, D. R., 128, *149*

Christou, A., 137, 138, *149*
Ciolkosz, T., 6, *151*
Clark, D. S., 127, *160*
Clator, I. G., 122, *149*
Clendenen, R. L., 169, 202, 207, 210, 214, 215, 216, 221, 230, *244*
Clifton, R. J., 25, 84, *149*
Clougherty, E. V., 208, *245*
Cohen, J. B., 134, 135, 140, *148, 149, 158*
Cohen, M., 208, *244*
Cohn, W. M., 168, *244*
Coleburn, N. L., 100, 109, 118, *149, 150*
Coleman, B. D., 24, *150*
Conze, H., 111, *150*
Cottrell, M. M., 23, *150*
Courant, R., 37, 39, *150*
Cowan, G. R., 26, *150*, 202, 203, 211, 242, *243*, 254, *279*
Cowin, S. C., 21, *150*
Cowperthwaite, M., 7, 35, 73, 79, 80, 81, 91, 93, 94, *150, 156, 160*
Crewdson, R., 23, 141, 142, 143, *147*
Crosnier, J., 76, 77, 110, 111, *150*
Curran, D. R., 21, 122, 130, 141, 143, *147, 150, 161*
Cutchen, J. T., 54, 103, *149, 150*

D

Daga, R., 135, *158*
Dapoigny, J., 89, *150*
Darnell, A. J., 98, *150*, 193, 195, *244*
Davies, G., 218, 231, *243*
Davis, B. L., 169, 192, 198, *243, 244*
Davis, R. O., 144, *150*
Davis, R. O., Jr, 22, 23, *150*
Davison, L., 94, *149*
Deal, W. F., 51, *150*
De Angelis, R. J., 134, 135, 140, *148, 149*

De Carli, P. S., 18, 98, 116, 118, *150, 157, 162*, 251, 255, *280*
Decker, D. L., 145, *150*, 181, 184, *244*
De Lai, A. J., 278, *280*
Deryagin, B. V., 250, *280*
Desoyer, J. C., 110, *155*
Dieter, G. E., Jr, 125, *150*
Dmitriev, M. E., 278, 279, *281*
Doran, D. G., 51, 78, 89, 97, 98, 100, 101, 102, 103, 104, 105, 106, 110, 113, 115, 116, 131, 138, *150, 151, 157, 159*
Dorman, J., 17, *157*
Drakin, V. P., 80, 117, 118, *147, 159*
Dremin, A. N., 80, 97, 105, 115, 117, 118, 119, *146, 151, 162*
Drickamer, H. G., 58, 98, 116, 145, *147, 151, 158, 159, 160*, 168, 169, 181, 182, 184, 198, 202, 203, 207, 210, 213, 214, 215, 216, 221, 230, 239, 240, 241, *243, 244, 246*, 253, 254, *280*
Drumheller, D. S., 22, 23, 48, 49, *157*
Dudoladov, I. P., 120, *147*
Duff, R. E., 120, *151*
Dugdale, J. S., *151*
Dulin, I. N., 118, *151*
Dunnebier, F., 17, *157*
Duvall, G. E., 2, 3, 5, 6, 18, 30, 34, 35, 39, 42, 51, 78, 91, 98, 100, 116, 124, 126, 129, *147, 151, 153, 161*
Duwez, P., 44, *151*

E

Eden, G., 18, 78, *151*
Edington, J. W., 135, *156*
Edwards, D. J., 80, *155*
Egorov, L. A., 89, *151*
Eichelberger, R. J., 58, 77, 78, 104, *151*
Erkman, J. O., 18, 130, *147, 151, 160*

Erlich, D. C., 14, 43, *161*
Eshelby, J. D., 107, *151*
Evans, L., 89, 90, *155, 158*
Evans, T., 274, *280*
Eversole, W. G., 250, *280*
Ewing, M., 17, *157*

F

Falicov, L. M., 99, *151*
Farber, J., 6, *151*
Farquhar, M. C. M., 208, *244*
Fedoseev, D. V., 250, *280*
Feklicher, E. M., 278, *280*
Ferman, M. A., 141, *155*
Foltz, J. V., 25, 132, 133, 139, *151, 159*
Forbes, J. W., 100, 109, 118, *149, 150*
Forgren, K. F., 168, *246*
Fowler, C. M., 117, 133, *157*, 202, 203, 210, 211, 212, 213, *244, 247*
Fowles, G. R., 2, 3, 5, 6, 9, 10, 18, 26, 30, 34, 35, 39, 42, 51, 78, 91, 92, 93, 94, 98, 100, 103, 116, 125, 126, 127, 129, 130, *147, 151, 152*
Fowles, R., 73, 74, 75, *162, 163*
Frasier, J. T., 80, *152*
Freeman, J. R., 54, *152*
French, B. M., 136, *152*
Freud, P. J., 169, *244, 245*
Frevel, L. K., 168, *245*
Friedrichs, K., 37, 39, *150*
Fritz, J. N., 8, 14, 22, 125, 129, 145, *152, 158*, 181, 182, 183, 184, 192, 196, *245*
Fritz, T. C., 191, 241, *245*
Fritzsche, J., 99, *152*
Froula, N. H., 128, *149*
Fuchs, R., 231, *245*
Fuller, P. J. A., 58, 59, 66, 67, 98, 131, *152*
Fumi, F. G., 182, *245*

Funtikov, A. I., 18, 130, *147, 156*
Fuyat, R. K., 187, 194, *247*

G

Gaffney, E. S., 115, *152*
Gager, W. B., 115, *152*
Garg, S. K., 23, *152*
Gatos, H. C., 98, *154*
Geballe, T. H., 111, *152*
George, J., 134, *152*
Gessaman, L. B., 51, *154*
Getting, I. C., 180, *245*
Gibson, R. E., 233, *243*
Gillis, P. P., 14, 15, 36, 126, 127, *152, 156*
Gilman, J. J., 11, 14, 15, 16, 125, *152*
Gilson, V. A., 85, *152*
Ginsberg, M. J., 71, 72, *152*
Gittings, E. F., 51, *152*
Godfrey, C., 67, *148*
Goranson, W., 51, *152*
Gordon, P., 134, 140, *148*
Goroff, I., 100, *152*
Gourley, L. E., 5, 123, *152, 154*
Gourley, M. F., 5, 123, *152, 154*
Grace, F. I., 25, 132, 134, 138, 139, 140, *151, 152, 159, 160*
Grace, J. D., 168, *244*
Grady, D. E., 59, 94, 124, *153*
Graham, R. A., 35, 52, 53, 54, 55, 56, 57, 94, 100, 102, 103, 104, 116, 122, 123, 125, 129, 130, 132, 144, *149, 153, 154, 155, 158, 159, 160, 161*
Green, A. E., 25, *153*
Greenberg, J. M., 24, *150, 153*
Gregson, V. G., 18, *146*
Grover, R., 144, 146, *153*
Grüneisen, E., 7, *153*
Gupta, Y. M., 127, 129, *147*

Gurtin, M. E., 24, 25, *149, 150, 153*
Gust, W. H., 117, 119, 120, 125, *146, 151, 153*
Gutmanas, E. Yu, 107, *162*

H

Hajj, F., 186, 187, 190, *245*
Hall, H. T., 169, 180, 181, 184, 224, 240, *243, 244, 245,* 258, 278, *280*
Hallouin, M., 110, *155*
Halpin, W. J., 18, 52, 53, 54, 55, 56, 102, 104, *153, 155, 158*
Hamann, S. D., 97, 98, 100, 105, 113, 116, 118, *154*
Hammond, D. E., 185, 187, 188, 190, 191, 192, *245*
Hanagud, S. V., 126, *147*
Hanneman, R. E., 98, 110, *154,* 254, 259, *280*
Harmon, T. C., 112, *162*
Haussühl, S., 181, *245*
Hauver, G. E., 58, 77, 78, 104, 105, *151, 154,* 181, *245*
Haygarth, J. C., 180, *245*
Hearst, J. R., 51, *154*
Heckel, R. W., 140, *154*
Herrera, R., 24, *150*
Herring, C., 111, *154*
Herrmann, W., 18, 19, 24, 25, 44, 45, 79, 131, *148, 154*
Heyda, J. F., 18, *154*
Hicks, D. L., 18, 45, *149, 154*
Higgens, G. T., 134, *154*
Hinze, E., 198, *246*
Hoffmann, R., 18, *154*
Hoge, K. G., 126, *152*
Holdridge, D. B., 18, *149*
Holland, J. R., 100, 122, 123, 125, 126, 127, 128, 132, 144, *153, 154, 155, 160*
Hollenbach, R. E., 46, 82, 83, 85, *147, 148*

Holt, A. C., 21, 125, *149, 153*
Hoover, W. G., 120, *151*
Hopkins, A. K., 22, *154*
Horie, Y., 121, 127, *149, 154*
Horiguchi, Y., 138, *154*
Horne, D. E., 141, *149*
Huang, K., 227, *244*
Huang, Y. K., 8, *154*
Hughes, D. S., 5, *154*
Hull, G. W., 111, *152*
Huston, E. E., 51, *152*

I

Ida, Y., 237, *245*
Ingram, G. E., 52, 54, 56, 57, 89, 125, *153, 154, 161*
Inman, M. C., 133, 134, 138, 139, *152, 160*
Irani, G. B., 51, *154*
Ishbell, W. M., 5, 18, *155, 160*
Ivanov, A. E., 105, 106, *155*
Ivanov, A. G., 104, 105, 106, 107, 108, 109, 126, *154, 155, 158, 159, 162*

J

Jacobs, R. B., 165, 168, 192, 198, *245*
Jacobs, S. J., 80, *155*
Jacquesson, J., 76, 77, 110, 111, *150, 155, 158*
Jajosky, B. C., Jr, 141, *155*
Jamieson, J. C., 118, *150,* 166, 168, 169, 198, 199, 200, *245, 246,* 251, 255, *280*
Jeffery, R. N., 240, *245*
Jenrette, B. D., 141, *149*
Jessup, R. S., 251, *280*
Johnson, J. N., 11, 15, 19, 35, 46, 47, 82, 83, 84, 116, 125, 126, 129, 130, *155*
Johnson, J. O., 18, *155*

Johnson, J. W., 187, *245*
Johnson, P. M., 85, *155*
Johnson, Q., 89, 90, 121, *155, 158,* 195, *245*
Johnston, W. G., 14, *155*
Jones, A. H., 5, *155*
Jones, D. E., 144, *153*
Jones, E. D., 6, *155*
Jones, G. A., 52, 53, *155*
Jones, H. D., 100, 109, *150*
Jones, O. E., 5, 11, 15, 52, 53, 54, 55, 56, 57, 100, 103, 124, 125, 126, 127, 128, 129, *153, 155, 158, 160, 161*
Jones, O. T., 11, 15, 46, 125, 129, *155*
Jones, W. H., 115, *152*

K
Kaechele, L. E., 142, *155*
Kaiser, W., 274, *280*
Kalashnikov, Ya. A., 278, *280*
Karnes, C. H., 18, 125, 126, *148, 149, 156*
Karpov, B. G., 80, *153*
Kasper, J. S., 168, 198, 200, *245, 246,* 252, 254, *280*
Katzman, H., 278, *280*
Kaufman, L., 208, *244, 245*
Keeler, R. N., 5, 51, 89, 98, 99, 103, 120, 123, *151, 155, 156, 158*
Kelly, J. M., 15, 25, 36, *156*
Kelly, P. M., 133, *148*
Kennedy, G. C., 145, 146, *162,* 166, 180, 190, 191, 193, 195, *245, 247*
Kennedy, J. D., 112, 125, *156*
Keough, D. D., 58, 59, 60, 62, 65, 66, 67, 69, 71, 94, 142, *148, 156, 162*
Kieffer, J., 89, *150*
Kirillov, G. A., 91, 100, 113, 114, 122, *156*
Kirsch, J. W., 23, *152*

Klein, A., 67, *148*
Klein, M. J., 115, 135, 140, *152, 156*
Kleinman, L., 100, *152*
Klement, W., Jr, 213, *245*
Kleschkynikov, O. A., 104, *159*
Knopoff, L., 144, *156, 161*
Kohn, W., 98, *156*
Kolesnikova, A. N., 18, *156*
Kologrivov, V. N., 110, 111, *160*
Kompaneets, A. S., 144, *156*
Kormer, S. B., 18, 91, 100, 113, 114, 115, 121, 122, 130, *147, 156, 157, 163*
Kormer, S. V., 203, 205, *243*
Kovach, R., 17, *157*
Kovenskii, I. I., 263, *280*
Kressel, H., 137, *157*
Krishkevich, G. V., 115, 121, *156*
Krupnikov, K. K., 6, 18, 121, *146, 148, 157*
Krupnikova, V. P., 18, *148, 157*
Kuleshova, L. V., 100, 101, 145, *147, 157,* 181, *243*
Kuriapin, A. I., 91, *163*
Kuryapin, A. I., 114, *157*
Kusubov, A. S., 59, 62, 66, 67, 131, *157*
Kuzin, N. N., 213, 240, *247*
Kuzubov, A., 192, *245*

L
La Mori, P. N., 169, 180, *244, 245*
Landeen, S. A., 51, *152*
Lapshin, A. I., 138, *148*
Larson, D. B., 128, 145, *157*
Latham, G., 17, *157*
Lawrence, R. J., 18, 44, 45, *154*
Lawson, A. W., 166, 168, *245*
Lee, E. H., 25, *157*
Lee, L. M., 18, *157*
Leibfried, G., 183, *245*
Libby, W. F., 98, *150,* 278, *280*

Lidiard, A. B., 107, *151*
Lieberman, R. C., 227, *243*
Liebermann, P., 18, *157*
Liebermann, R., 220, 221, 234, *245*
Lighthill, M. J., 39, *157*
Linde, R. K., 5, 14, 17, 18, 20, 21, 43, 47, 55, 73, 97, 98, 100, 102, 103, 105, 106, 107, 113, 115, 116, 131, 138, *151, 157, 161, 162*
Lippert, L., 194, *247*
Lipson, H., 208, *244*
Lisitsyn, Yu. V., 105, 106, 107, 108, 109, *155, 158, 162*
Liu, D. T., 25, *157*
Liu, L. G., 184, 203, 204, 206, 207, 216, 217, 220, 222, 225, 226, 233, *245, 247*
Lombardi, M., 18, *159*
Loree, T. R., 117, 133, *157*
Ludwig, W., 183, *245*
Luedmann, H. D., 180, *245*
Lundergan, C. D., 22, 23, 48, 49, 79, 131, *148, 157*
Lutze, A., 6, *147, 157*
Lyle, J. W., 65, 66, 67, 125, *158*
Lynch, R. W., 169, *244*
Lysne, P. C., 18, 22, 35, 56, 57, 89, 104, 116, *158*
Lyuttsau, V. G., 250, *280*

M

Ma, C. H., 135, *158*
Ma, Ching H., 128, *158*
McCollum, W. A., 193, 195, *244*
MacDonald, D. K. C., *151*
McKinstry, H. A., 181, *246*
McQueen, R. G., 2, 8, 14, 22, 30, 35, 51, 90, 91, 116, 119, 125, 126, 129, 145, *152, 158, 160, 162*, 166, 181, 182, 183, 184, 192, 196, 203, 205, 231, *245, 246*
McSkimin, H. J., 166, *246*

McWhan, D. B., 168, 184, 217, *246*
Madelung, E., 231, *245*
Mader, C. L., 86, 89, 141, *148*
Mahajan, S., 132, *158*
Maiden, C. J., 5, *155*
Majumdar, A. J., 197, *245*
Malvern, L. E., 13, 45, *158*
Manghnani, M. H., 231, *245*
Mao, H., 145, *148*
Mao, H. K., 179, 192, 194, 195, 196, 201, 203, 204, 206, 207, 208, 209, 212, 213, 214, 215, 216, 217, 219, 220, 221, 222, 223, 224, 230, 232, 235, 236, 240, *244, 245, 246, 247*
Marsh, D. J., 132, *158*
Marsh, S. P., 8, 14, 22, 90, 119, 125, 129, 145, *152, 158*, 181, 182, 183, 184, 192, 196, 203, 205, 231, *245, 246*
Martin, J. E., 168, *246*
Mason, D. S., 18, 44, 45, *154*
Maxwell, D. E., 18, *154*
Mayer, J. E., 186, 187, *246*
Meissner, R., 17, *157*
Melani, A., 181, *245*
Merrill, L., 181, 184, *244*
Messmer, R. P., 275, *280*
Meyer, H. O. A., 273, *280*
Michaels, T. E., 11, 15, 46, 125, 129, *155*
Migault, A., 76, 77, 110, 111, *150, 158*
Mikkola, D. E., 135, 140, *158*
Miles, M. H., 126, 129, *147*
Milledge, H. J., 273, *280*
Miller, R. A., 206, *246*
Mineev, V. N., 104, 105, 106, 107, 108, 109, *154, 155, 158, 159, 161, 162*
Minomura, S., 145, *158*
Minshall, F. S., 98, 115, 117, *147, 157*, 202, 203, 207, 210, 211, 212, 213, 239, *243, 244, 247*

Mitchell, A. C., 89, 90, 120, 121, 123, *151*, *155*, *158*
Mitchell, A. D., 98, 99, 123, *156*
Mitchell, D. W., 135, *158*
Mochalov, S. M., 122, *148*
Modenov, V. P., 278, 279, *281*
Moiseev, B. N., 120, *161*
Moore, M. J., 198, 200, *246*
Morland, L. W., 9, 130, *158*
Mote, J. D., 11, 15, 54, 55, 125, 126, 129, *155*
Munro, D. C., 168, *244*
Munson, D. E., 22, 79, 125, 126, 127, *149*, *158*
Muranevich, A. K., 118, *146*
Murnaghan, F. D., 7, *158*, 187, 188, *246*
Murr, L. E., 132, 133, 134, 138, 139, *152*, *159*
Murri, W. J., 79, 80, 81, 94, 96, 100, 101, 105, 106, 125, 126, 129, *147*, *151*, *157*, *159*, *160*
Myers, W. R., 169, *244*

N

Nafe, J. E., 226, 227, 228, 234, *243*
Naghdi, P. M., 25, *153*
Nakamura, Y., 17, *157*
Naumann, R. J., 122, *159*
Neilson, F. W., 52, 53, 54, 55, 56, 57, 102, 103, 124, 125, 126, *153*, *155*, *159*
Nelson, A., 134, 135, *149*
Neuhaus, A., 198, *246*
Newey, C. W. A., 107, *151*
Ng, R., 18, *159*
Nikol'skaya, I. V., 278, *280*
Nitochkina, E. V., 89, *151*
Noll, W., 24, *160*
Nomura, Y., 138, *159*
Novikov, S. A., 126, *154*, *159*

Novitskii, E. Z., 104, 105, 106, 108, 109, *154*, *155*, *158*, *159*, *162*
Nunziato, J. W., 24, *154*

O

O'Brien, J. F., 57, *159*
Orava, R. N., 139, *159*
Orekin, Yu. K., 89, *151*
Orlov, Yu. L., 278, *280*
O'Rourk, J., 202, 203, 210, 211, 212, *247*
Orowan, E., 128, *159*
Owen, N. B., 168, *246*

P

Palmer, E. P., 110, *159*
Pastine, D. J., 18, *159*
Paul, W., 99, *159*
Pavlovskii, M. N., 80, 100, 117, 118, 119, 145, *147*, *159*, 181, *243*
Peck, J. C., 23, *159*
Peltzer, C. P., 18, *147*
Percival, C. M., 56, *158*
Perez-Albuerne, E. A., 145, *159*, 168, 169, 181, 182, 184, *244*, *246*
Perls, T. A., 73, *159*
Perry, F. C., 6, *160*
Pershin, S. V., 80, 117, 118, 119, *146*, *151*
Petersen, C. F., 18, 79, 80, 81, 96, *159*, *160*
Peterson, E. L., 98, 115, *147*, 207, 212, 239, *243*
Peyre, C., 126, *160*
Piermarini, G. J., 169, 192, 195, *246*, *247*
Pistorius, C. W. F., 194, 195, *246*
Plyukhin, V. S., 110, 111, *160*
Podurets, M. A., 120, *161*
Pogorelov, V. F., 80, 117, *151*
Pokhil, P. F., 80, *163*
Popov, L. V., 120, *161*

Popova, L. T., 100, 113, *156*
Power, D. V., *154*
Pratt, P. L., 107, *151*
Preban, A. G., 134, 140, *148*
Press, F., 17, *157*
Price, J. H., 58, 59, 66, 67, 98, 131, *152*
Prindle, H., 18, *160*
Pujol, J., 126, *160*

R

Rayleigh, Lord, 36, *160*
Rempel, J. R., 18, *160*
Rice, M., 125, 126, *161*
Rice, M. H., 2, 8, 30, 35, 51, 91, 116, 121, 145, *160, 162,* 166, *246*
Richtmeyer, R. D., 40, *162*
Riggleman, B. M., 198, *246*
Riley, N. A., 168, *245*
Ringwood, A. E., 118, *146,* 231, *243*
Rogers, F. J., 144, *153*
Rohde, R. W., 55, 57, 103, 117, 125, 127, 129, 132, *155, 160*
Romain, J. P., 110, *155*
Romanova, V. I., 144, *156*
Rose, M. F., 122, 132, 133, 135, 136, 138, *149, 159, 160*
Rosenberg, J. T., 14, 43, *161*
Ross, M., 120, *151*
Rossi, L., 234, *246*
Rossini, F. D., 251, *280*
Rotter, C. A., 204, *246*
Roy, R., 197, *245*
Royce, E. B., 5, 51, 98, 99, 103, 117, 119, 120, 122, 123, 124, 125, 128, *146, 151, 153, 156, 160, 161, 162*
Rozanov, O. K., 105, *162*
Ruderman, M. H., 86, *146*
Rudman, P. S., 140, *156*
Russell, T. L., 127, *160*
Ryabor, V. A., 250, *280*

S

Sadkov, Yu. A., 213, 240, *247*
Samara, G. A., 35, 58, 116, *158, 160*
Sammis, C. G., 227, *246*
Samylov, S. V., 90, 111, *149*
Savage, R. H., 277, *280*
Schall, R., 89, *160*
Schmidt, D. N., 5, 18, 55, 73, *157, 160*
Schock, R. N., 198, 199, 200, *246*
Schreiber, E., 220, 221, 227, 230, 232, 234, *243, 245, 246*
Schriever, R. L., 65, 66, 67, *158*
Schuele, D. E., 181, 206, *243, 246*
Schuler, K. W., 22, 25, 46, *158, 160*
Sclar, C. B., 169, *245*
Seaman, L., 14, 17, 18, 20, 21, 23, 40, 43, 47, 48, 141, 152, 143, *147, 157, 160, 161*
Seay, G. E., 123, *161*
Semerchan, A. A., 213, 240, *247,* 278, 279, *281*
Shaner, J. W., 123, *161*
Shapiro, J. N., 144, *156, 161*
Shepard, M. L., 184, 207, *246*
Shimmin, W., 67, *148*
Shockey, D. A., 141, 143, *161*
Short, N. M., 136, *152*
Shunk, R., 77, *148*
Shvedov, K. K., 80, *151, 163*
Silberman, M. L., 202, 204, 208, 209, 210, 211, 212, 215, 217, *246*
Simakov, G. V., 120, *161*
Simon, F., 252, *279*
Simonenko, V. A., 35, *163*
Sinitsyn, M. V., 91, 100, 113, 114, 121, 122, *156, 157, 163*
Sinitsyn, V. A., 126, *154, 159*
Skidmore, I. C., 141, *161*
Skoog, C., 67, *148*
Slagle, O. D., 181, *246*
Slater, J. C., 8, *161*

Slesarev, V. N., 278, *280*
Slykhouse, T. E., 198, *246*
Smith, C. P. M., 18, *151*
Smith, C. S., 204, *246*
Smith, C. W., 94, 96, *156*, *159*
Smith, J. F., 184, 207, *246*
Smith, J. H., 141, *161*
Smith, P. L., 168, *246*
Soga, N., 184, 217, 222, 226, 230, 233, *243*, *246*, *247*
Speranskaya, M. P., 130, *147*
Stanko, W. S., 250, *279*
Sterenberg, L. E., 278, *280*
Stevens, A. L., 5, *161*
Stevens, D. R., 277, *280*
Stirpe, D., 117, *157*
Stook, P. W., 169, 171, 224, *244*
Stromberg, H. D., 277, *280*
Strong, H. M., 58, 110, *148*, *154*, 250, 252, 254, 258, 259, 260, 261, 264, 272, 273, 274, *280*
Styris, D. L., 98, *161*
Sullivan, B. R., 80, *146*
Sutton, G., 17, *157*
Sutulov, Yu. N., 120, *147*
Swanson, H. E., 187, 194, *247*
Swenson, C. A., 97, *161*
Swift, R. P., 73, 74, 75, *162*, *163*
Syono, Y., 237, *245*

T

Taborsky, L., 139, *161*
Takahashi, T., 145, *148*, 169, 171, 173, 176, 179, 180, 181, 182, 183, 184, 192, 193, 194, 195, 196, 197, 198, 199, 200, 201, 203, 204, 206, 207, 208, 209, 212, 213, 214, 215, 216, 217, 218, 219, 220, 221, 222, 223, 224, 225, 226, 230, 231, 232, 233, 235, 236, 240, *243*, *244*, *245*, *246*, *247*
Tang, T. Y., 168, *245*

Thomsen, L., 182, 183, 184, 188, *247*
Taylor, G. I., 36, *161*
Taylor, J. W., 8, 11, 14, 15, 22, 45, 90, 125, 126, 128, 129, *152*, *158*, *161*
Taylor, S. M., Jr, 22, *151*
Tchen, W., 18, *159*
Tetelman, A. S., 142, *155*
Thiel, M., von, 59, 62, 66, 67, 131, *157*
Thouvenin, J., 18, 126, *160*, *161*
Thurnborg, S., *161*
Thurston, R. N., 191, 241, *245*
Tinder, R. F., 126, 129, *147*
Tobisawa, S., 138, *159*
Toskoz, N., 17, *157*
Torvik, P. J., 22, *161*
Tosi, M. P., 182, 187, *245*, *247*
Towne, T. L., 57, 125, *160*, *161*
Trueb, L. F., 134, 135, 136, *161*, 254, 276, *280*
Truesdell, C., 24, *161*
Trunin, R. F., 6, 120, 121, *146*, *161*, 203, 205, *243*
Tsou, F. K., 22, *161*
Tuler, F. R., 141, *161*
Turner, G. H., 110, *159*
Tyunyaev, Yu. N., 105, 106, 107, 108, 109, *155*, *158*, *161*, *162*

U

Ugrinic, G. M., 187, *247*
Urlin, V. D., 18, 91, 113, 114, 121, 122, *156*, *162*, *163*
Urusovskaya, A. A., 106, 107, *162*

V

Vaidya, S. N., 145, 146, *162*, 166, 190, 191, 193, 195, *247*
Valitskii, V. P., 122, *148*
Vanfleet, H. B., 180, 240, *245*
Van Valkenburg, A., 198, 199, *247*
Varfolmeeva, T. D., 278, *280*
Vashchenko, V. Y., 118, *151*

Vasilyev, L. V., 126, *159*
Venable, D., 86, 89, 141, *148, 162*
Vereshchagin, L. F., 168, 213, 240, *247*, 278, 279, *280, 281*
Verma, R. K., 232, 233, *247*
Villere, M. P., 136, *160*
Vladimirov, L. A., 130, *147*
Vodar, B., 89, *150*
von Mises, R., 6, 39, *162*
von Neumann, J., 40, *162*
Vydyanath, H. R., 133, 134, *159*

W

Wackerle, J., 18, 125, 126, *155, 162*
Wagner, G., 194, *247*
Wakatsuki, Masa, 265, *281*
Walsh, E. K., 25, *160*
Walsh, J. M., 2, 8, 30, 35, 51, 89, 91, 116, 121, 144, 145, *160, 162*, 166, *246*
Wang, A. S. D., 22, *149*
Warnes, R. H., 117, 126, 133, *157, 162*
Warnicka, R. L., 141, *162*
Warren, B. E., 140, *162*
Warren, N., 21, *162*
Warshauer, D. M., 99, *159*
Wasley, R. J., 57, 126, *152, 159*
Watkins, G. D., 275, *280*
Wayne, R. C., 123, *161, 162*
Weaver, J. S., 145, *148*, 179, 180, 181, 182, 183, 184, 192, 194, 195, 196, 219, 221, 223, 227, 230, 232, *244, 246, 247*
Weidner, D. J., 128, *162*
Weill, A. R., 208, *244*
Weir, C. E., 169, 192, 193, 195, *246, 247*
Weiss, R. J., 208, *245*
Wentorf, R. H., 118, *148*, 250, 252, 256, 257, 258, 260, 263, 265, 268, 273, 274, 277, *280, 281*
White, R. M., 107, *162*

Whitworth, R. W., 108, *162*
Wierzbicki, T., 25, *157*
Wild, R. K., 274, *280*
Will, H. A., 250, *279*
Willardson, R. K., 112, *162*
Williams, E., 67, *148*
Williams, E. O., 73, *162*
Williams, R. F., 14, 26, 43, 58, 59, 67, 71, 73, 91, 92, 93, 94, *150, 152, 156, 161, 162*
Wong, J. Y., 65, 67, 98, 99, 107, 116, 123, *156, 162*
Wood, D. S., 127, *160*
Worlton, T. G., 169, *244*
Wright, A. J., 168, *246*
Wright, L. D., 123, *161*
Wright, P. W., 78, *151*
Wu, J. H., 22, *150*

Y

Yakovlev, E. N., 278, *280*
Yakushev, V. V., 105, 108, 115, *162*
Yakusheva, O. B., 115, *162*
Yampol'skii, P. A., 144, *156*
Yanov, V. A., 105, 106, *154*
Yarger, F. L., 8, *162*
Yates, M., 17, *157*
Young, C., 73, 74, 75, *162, 163*
Young, E. G., 45, *154*
Youngblood, J. L., 140, *154*
Yushko, K. B., 115, *156*
Yushko, K. R., 121, *163*

Z

Zababakhin, E. I., 35, *163*
Zaitsev, V. M., 80, *163*
Zel'dovich, Ya. B., 91, 105, 114, 121, *163*
Zlatin, N. A., 122, *148*
Zubarev, V. N., 118, *151*
Zukas, E. G., 117, 133, 139, *157, 163*, 202, 203, 210, 211, 212, 213, *244, 247*

Subject Index

A

Alkali
and alkaline earth metals in shock
waves, 18, 120, 122, 125, 128,
129, 131, 135, 143
halides, other salts in shock waves,
18, 46, 55, 58, 91, 100, 101,
104–108, 113, 115, 117, 122,
125, 128, 129, 138, 140, 145
halides in static compression, 192–
197, 241
halides
melting of at pressure, 195
see also Melting
halides
phase transformations in, 192–
197
Aluminium (aluminum) and its
alloys in shock waves, 18, 43,
68, 89, 122, 125, 126, 130,
131, 135, 139, 142, 143, 145
Anisotropy
effect on shock waves, 129, 130
on diamond anvils, 177, 199,
242–243
in compressibility of substances,
see Axial ratio
Antimony in shock waves, 117, 125
Artificial viscosity method in shock
waves, 40, 42–43, 47, 49
Attenuation of shock waves, 16, 25,
30, 130
Axial ratio, c/a
in ilmenite, 238–239
in metals, 213–218, 242
in stishovite, 236–238

B

Ballas diamond, 276
Bauschinger effect, 43–44, 127–128
Beryllium in shock waves, 18, 125,
128, 129, 143
Birefringence in diamonds, 173
Bonding together of diamonds, 276–
279
Boron
in diamonds, 273
nitride in shock waves, 18, 90,
101, 118
Bridgman anvils, 168, 198
Bulk modulus of compression versus
composition, 226–235

C

Carbides
at high pressure, 259–261, 277
in shock waves, 140
Carbon
in shock waves, 18, 70–72, 118–
119, 135, 136
phase diagram, 251
solvents, 257, 264, 269
Carbonaceous materials to diamond, 256, 262
Carbonado diamond, 276
Catalysts for diamond growth, 259

Cesium iodide as pressure reference substance, 184–192
P–V relation, 184–192
Cleavage of diamond, 276
Composite materials in shock compression, 22–24, 48–49
Copper in shock waves, 18, 46, 76, 77, 90, 110, 111, 120, 125, 126, 129, 130, 134, 135, 140, 143
Critical growth rates for diamond, 261, 272

D

Deviator stress, 9, 35, 67
Diamond
    anvils, 169–178, 198, 242–243
    by shock compression of graphite, 135, 136, 254, 255
    crystal structures (cubic, hexagonal), 252, 266
    from crystalline graphite, 253, 255, 265–267
    from solid carbons, 255, 256, 257
    — graphite equilibrium, 252
    growth
        in electric fields, 264
        in temperature gradient, 268
        on seed crystals, 269
        oriented by parent graphite, 265–267
    synthesis at low pressures, 250
Diamonds
    colour in, 273
    optical properties, 273–275
Dielectric breakdown in shock waves, 102–104
Diffusion of carbon in molten metals, 261
Dislocations in shocked solids, 14–16, 45–46, 50, 125–129, 133–135, 137, 139

E

Elastic–plastic solids, 8–11, 41–44
Elastic precursor in shocked materials, 35, 42, 45, 46, 125, 126
Electrets in shock waves, 108
Electric polarization in shock waves, 101–108
Electromagnetic transducers, 73, 80
Electronic structure of shocked metals, 11, 120
Entropy changes in phase transformations of alkali halides, 194, 196, 241
Equation of state of shocked solids, 7, 8, 50
Eulerian and Lagrangian coordinates, 26–28, 36–38, 182, 188

F

Ferroelectric materials in shock waves, 103–104
Flyer plate, 3–6, 14, 33, 45, 58, 59, 142
Free surface velocity, 29, 78–79
    measurement of, 76

G

Garnets
    compression of, 184, 218, 220, 222, 225, 226, 233, 235
Germanium in shock waves, 100, 108, 110, 112, 119, 125
Graphite
    alteration by compression, 253–254
    crystal structure, 252, 254, 266
    from solutions of carbon, 258–259, 260
    intercalation and diamond formation, 267
Growth layers on diamonds, 270, 271

H

Hall effect in shock waves, 112

Haematite

compression of, 177, 219, 221

Hexagonal diamond, 252, 253

Hugoniot

elastic limit (HEL), 119, 124, 126–129, 144

relations for shock waves, 29, 31–36, 50, 93

I

Immersed foils in shock waves, 86

Impact techniques, 5, 6

Impedance matching in shock waves, 32–34, 56, 57

Impurities in diamond, 260, 272

In-material gauges for shock waves, 28, 36, 38, 58, 65, 67, 68, 71, 73, 79–82, 86, 92–97

Interferometer techniques in shock waves, 82

Iron and iron alloys

in shock waves, 18, 47–49, 93, 116–117, 120, 122–123, 126, 127, 128, 129, 130, 132, 133, 136–138, 143, 202, 203, 210, 211, 212, 213

in static compression, 201–205, 213–216, 242

phase changes in, 207–213, 242

Iron–carbon phase diagram, 260

L

Lagrangian

and Eulerian coordinates, 26–28, 36–38, 182, 188

gauges in shock waves, *see* In-material gauges

Lasers

use of in shock waves, 2, 6, 57, 82–86

Lattice defects and shock waves, 137

Lead

compression of, 203–206, 217

phase changes in, 213, 242

Light emission in shock waves, 112–114

see also temperatures in shock waves

M

Magnesium oxide

in shock waves, 100, 115, 118, 140

$P–V$ relations, 183, 230

Magnetic effects from shock waves, 122, 136

Magnetite

phase changes in, 235–236

static compression of, 219, 222, 232

Manganin pressure gauges, 48

in shock waves, 58–60, 65–69

Melting

in shock waves, 91, 101, 115, 121–122

of alkali halides under pressure, 195

of carbon, 255

Memory effects in solids, 24

Method of characteristics for shock waves, 39–40, 42, 45, 49

Microstructures of shocked solids, 132, 138–141

Momentum traps, 131

N

Natural diamonds compared with synthesized diamonds, 279

Nitrogen in diamonds, 273

Nickel and nickel alloys in shock waves, 99, 111, 117, 122, 123, 132–134, 138, 139

Nickel–carbon phase diagram, 260

O

Orientation
  of diamond from parent graphite,
    265–267
  of metal films on diamond, 262
Oxygen on diamond surfaces, 276
Oxides
  in shock waves, 18, 51–55, 65, 81,
    89, 100, 102–104, 108, 115,
    118, 125, 126, 133, 140
  static compression of, 218–235

P

Phase changes in shock waves, 35, 71,
    72, 89, 90, 105, 110, 115–122,
    192
Phase transformations
  in alkali halides, 192–197
  in iron and its alloys, 207–213, 240
  in lead, 203, 240
  in magnetite, 235–236
  in silver halides, 197–200, 241
    as pressure reference points,
      239–243
Piezoelectric transducers, 51
Piezoresistive transducers, 57
Polar solid transducers, 77
Polarization, see Electric polarization
Polymers and shock waves, 18, 46–49,
    56–57, 67, 68, 81, 91, 101,
    105, 143
Porous solids in shock compression,
    16–21, 47–48, 122
Power supplies for instrumentation,
    59–65, 69–72
Pressure gradients on diamond
    anvils, 172–176
P–V relations
  for various substances, see sub-
    stance in question
  from shock wave data, 7–8, 143–
    144, 166–167

Q

Quartz pressure gauges, 2, 52, 102,
    104

R

Rare earth metals in shock waves,
    70–72, 109, 120
Rayleigh line in shock compression,
    31–33
Relaxation in shocked solids, 10–13,
    44–46, 125
Reverbration plate in shock waves,
    54, 56–57
Rhenium
  bulk modulus of, 184
  compression of, 204, 206, 207, 214,
    217

S

Seismic parameter versus compo-
    sition, 228–235
Semiconductors in shock waves,
    108–109
  see also Germanium and Silicon
Semiconducting diamonds, 274, 275
Sharpening or steepening of shock
    fronts, 30, 39
Shock wave
  formation, 38
  propagation in various classes of
    solids, 41, 44, 46, 47
  schematic drawing, 4
Shock waves in fluids, 31
Silica, static compression of various
    forms, 220, 224–225, 231,
    237
Silicates
  in shock waves, 18, 51, 52, 81,
    118, 125, 136, 138
  static compression of, 218–235
Silicon in shock waves, 108, 109, 112,
    119

Silver halides, compression of, phase changes in, 197–200
Single crystals in shock waves, 46, 55, 125, 126, 129, 135
Sintering of diamonds, 277
Sodium chloride
  as pressure reference substance, 179–184
  in shock waves, 46, 55, 91, 100, 104–108, 113, 115, 122, 125, 128, 129, 145, 181, 192
  P–V relation, 179–184
Spinels, static compression of, 219, 220, 222–224, 232
Steady flow in shock waves, 28
Stishovite, static compression of, 177, 224–225, 231, 237
Strain waves, one-dimensional, 26
Strength of material, effect of on shock structure, 34
  see also Hugoniot elastic limit

T
Tantalum in shock waves, 125, 126
Temperatures in shock waves, 90–91, 111–114, 133, 134
Tetrahedral press for X-ray studies, 169, 177

Thermal
  conductivity of diamond, 274
  equilibrium in shock waves, 114
Thermoelectric effect in shock waves, 76–77, 109–112

U
Unsteady wave propagation, methods of solution, 39

V
Vaporization in shock waves, 122

W
Water in shock compression, 121
Work hardening and shock waves, 127, 138

X
X-rays, use of in shock waves, 86, 143

Y
Ytterbium in shock waves, 70–72, 120
Yttrium-iron garnet in shock waves, 124